Polyhedral and Algebraic Methods in Computational Geometry

Polyhedral and Algebraic Methods in Computational Geometry

Editor

Nik Pachis

scitus
academics

Polyhedral and Algebraic Methods in Computational Geometry
Edited by **Nik Pachis**

ISBN: 978-1-68117-449-5
Library of Congress Control Number: 2016937605

SCITUS Academics LLC,
www.scitusacademics.com
Box No. 4766, 616 Corporate Way,
Suite 2, Valley Cottage,
NY 10989

Notice

Printed in United States of America on Acid Free Paper

Preface

Computational geometry is a branch of computer science devoted to the study of algorithms which can be stated in terms of geometry. Some purely geometrical problems arise out of the study of computational geometric algorithms, and such problems are also considered to be part of computational geometry. While modern computational geometry is a recent development, it is one of the oldest fields of computing with history stretching back to antiquity.The main impetus for the development of computational geometry as a discipline was progress in computer graphics and computer-aided design and manufacturing (CAD/CAM), but many problems in computational geometry are classical in nature, and may come from mathematical visualization.

Other important applications of computational geometry include robotics, geographic information systems, integrated circuit design, computer-aided engineering, and computer vision. Polyhedral and Algebraic Methods in Computational Geometry provides a thorough introduction into algorithmic geometry and its applications. This book is directed towards advanced undergraduates in mathematics and computer science, as well as towards engineering students who are interested in the applications of computational geometry.

Table of Contents

CHAPTER 1

Checking Geometric Programs or Verification of Geometric Structures

Kurt Mehlhorn[a], Stefan Näher[b], Michael Seel[a,], Raimund Seidel[c], Thomas Schilz[a], Stefan Schirra[a], Christian Uhrig[d]

[a] *Max-Planck-Institute for Computer Science, Saarbrücken, Germany*

[b] *Fachbereich Mathematik und Informatik, Martin-Luther-Universität, Halle, Germany*

[c] *Fachbereich Informatik, Universität des Saarlandes, D-66040 Saarbrücken, Germany*

[d] *Algorithmic Solutions GmbH, Saarbrücken, Germany*

ABSTRACT

A program checker verifies that a particular program execution is correct. We give simple and efficient program checkers for some basic geometric tasks. We report about our experiences with program checking in the context of the LEDA system. We discuss program checking for data structures that have to rely on user-provided functions.

INTRODUCTION

Programming is a notoriously error-prone task; this is even true when programming is interpreted in a narrow sense: going from a (correct) algorithm to a program. The standard way to guard against coding errors is program testing. The program is exercised on inputs for which the output is known by other means, typically as the output of an alternative program for the same task. In geometric programming visual inspection of outputs is a powerful testing method, it is however limited to programs operating in low-dimensional space. Program testing has severe limitations:

- It is usually only done during the testing phase of a program. Also, it is difficult to determine the "correct" suite of test inputs.
- Even if appropriate test inputs are known it is usually difficult to determine the correct outputs for these inputs: alternative programs may have different input and output conventions or may be too inefficient to solve the test cases.

Given that program verification, i.e., formal proof of correctness of an implementation, will not be available on a practical scale for some years to come, *program checking* has been proposed as an extension to testing [2,3]. The cited papers explored program checking in the area of algebraic, numerical, and combinatorial computing. We explore it for geometric computing anData structures and we also report on first implementations within the LEDA system [11,14].

Consider a program P that is supposed to compute a function f. A program checker for f is a program C that takes a pair (x, y) and checks whether $y = f(x)$. The requirements for a program checker are the following.

- *Correctness*. The checker must be able to correctly decide whether y is equal to $f(x)$.
- *Simplicity*. The previous item suggests that we are in an infinite loop: in order to check a program we exercise a checker whose correctness we again have to check, ... The simplicity requirement breaks
- the loop. Checkers should be so simple that their correctness can be established beyond reasonable doubt. [1]
- *Effi* . The resource requirement of a program checker should be lower than the one of the checked program. An efficient program checker can be used in the field, i.e., every execution of the original program can be checked.

The current paper has three parts. In the first part we give program checkers for a number of problems iŋ computational geometry: convexity of polygons and polytopes, convex hulls in arbitrary dimensions, Voronoi anDelaunay diagrams and trapezoidal decompositions. We will see that program checking is sometimes facilitated by augmenting the program checked; for example a convex hull program that delivers a triangulation of the hull is much easier to check than a program that only returns the hull polytope.

In the second part we report about our practical experience with program checking within the LEDA system. We implemented most checkers described in this paper [10,12]. In particular, we added checking to a program that computes convex hulls in arbitrary dimensional space. The hull program is structured into four layers: arbitrary precision integer arithmetic, linear algebra over the rationals, geometric primitives for higher dimensional space, and finally the convex hull program. We made the last three layers self-checking. The addition of self-checking increased the running time by less than a factor of two, decreaseDebugging time considerable, and greatly increased our confidence that the implementation produces correct results.

Large software systems are built in layers. As the number of layers increases debugging becomes harder as errors become harder to locate. If all programs in a system are self-checking any program can safely assume that all the lower level programs used by it are correct. *We believe that program checking will play a prominent role in the construction of software libraries and have started to add checking facilities to LEDA. We also believe that programs in a library should be designed for check-ability.*

Can we make all programs in a library self-checking? There are two obstacles to that. Firstly, there might be problems where verification brings about an unbearable overhead. Secondly, programs in a library might have to rely on user provided functions over which the library has no control. The data structure area provides examples for this situation.

In the third part we consider tree-based implementations of the dictionary data type. A dictionary realizes a finite subset S of some key type K under the operations insert, delete, and member. For tree- based implementations the key type K must be linearly ordered. In LEDA the user of the dictionary data type provides a function *compare* that is supposed to compute a linear order on K. Can a dictionary implementation guard against an error in the compare-function? We will show that some tree-baseDictionary implementations can indeed guard against errors in the compare function and still maintain logarithmic search and update time. We will also argue that the efficiency of finger searches cannot be maintained in an environment where the compare may be faulty.

VERIFYING GEOMETRIC STRUCTURES

We give program checkers for some problems in computational geometry.

Directions and convex polygons

A direction is any non-zero vector in two-dimensional Euclidean space. For a direction d let $\alpha(d)$ denote the angle between the positive x-axis anD, see **Figure**. 1. Then, $0 \leq \alpha(d) < 2\pi$. Define a linear order on directions as follows: if d1 anD2 are directions then d1 ≤ d2 iff α(d1) ≤ α(d2). Let d0, d1,... , dn−1 be a sequence of directions. We call it counter-clockwise ordered (CCW-ordered) if there is an i such that 0 ≤ α(di) ••• < α(di−1)< 2π and we call it weakly counter-clockwise ordered (weakly CCW-ordered) if there is an i such that 0 ≤ α(di) ≤ α(di+1) ≤ ••• ≤ α(di−1)< 2π . In both definitions indices are to be interpreted modulo n. Intuitively, a cyclic sequence of directions is (weakly) CCW- ordered if it winds around the origin once.

It is easy to check whether a sequence d0,..., dn−1 is CCW-ordered. We simply scan the sequence once and check that all elements are distinct from their cyclic successor and that all but one element is smaller than its cyclic successor. It is equally simple to check for weakly CCW-orderedness.

We use the orderedness predicate for directions to check convexity of polygons. Let $v0$, $v1$, . . . , $vn{-}1$ be a cyclic sequence of points in the plane. It is called a counter-clockwise convex polygon (weakly counter-clockwise convex polygon, respectively) if consecutive points in the sequence are distinct and if the sequence $vi{+}1 - vi$, $0 \leq i \leq n-1$, of directions if CCW-ordered (weakly CCW-ordered, respectively).

Remark. If the points vi have rational coordinates then the vectors $vi{+}1 - vi$ have rational coordinates. Moreover, for rational directions $d1$ anD2 the ordering predicate $d1 \leq d2$ can be decided using rational arithmetic: determine the quadrants containing the directions first and if both directions belong to the same quadrant compare the slopes. Thus there is no need to compute trigonometric functions although such use is suggested by the definition of the ordering relation.

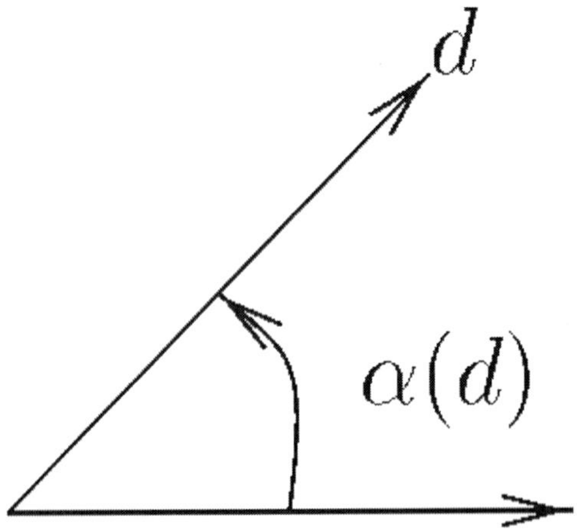

Figure 1. The angle *α(d)* defi ed by a direction *d*.

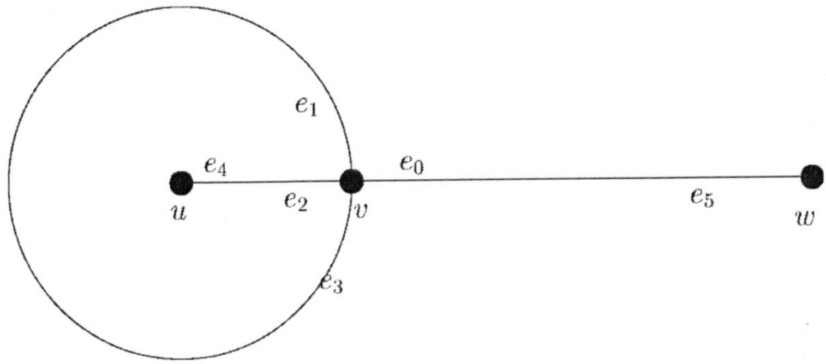

Figure 2. A map $M = (V, E)$ with six edges $e0 = (v, w)$, $e1 = (v, v)$, $e2 = (v, u)$, $e3 = (v, v)$, $e4 = (u, v)$, and $e5 = (w, v)$. We have $rev(e0) = e5$, $rev(e1) = e3$, and $rev(e2) = e4$. The cyclic ordering on the edges with source v is $e0$, $e1$, $e2$, $e3$, i.e., $as(e0) = e1$, $as(e1) = e2$, $as(e2) = e3$, and $as(e3) = e0$. Also, $as(e4) = e4$ and $as(e5) = e5$. The face cycle successors are as follows: $fs(e0) = e5$, $fs(e1) = e2$, $fs(e2) = e4$, $fs(e3) = e0$, $fs(e4) = e1$, and $fs(e5) = e3$. The face cycles of M are: $e0$, $e5$, $e1$ and $e2$, $e4$, $e1$. The genus of M is zero. The drawing is an embedding of M.

The embedding partitions the plane into two regions, one bounded and one unbounded. The unbounded region corresponds to the face cycle $e3$, $e0$, $e5$ and the bounded region corresponds to the face cycle $e1$, $e2$, $e4$.

Embedded graphs

Basics

We deal with combinatorially and geometrically embedded undirected graphs. When dealing with such graphs it is convenient to distinguish between the two orientations of an edge. We therefore phrase our discussion in terms of so-called bidirected graphs.

A bidirected graph is a directed graph $G = (V, E)$ together with a bijection $rev : E \rightarrow E$ such that for all edges e: $rev(e)$ 6= e and if $e = (v, w)$ then $rev(e) = (w, v)$ and $rev(rev(e)) = e$. The condition $rev(e)$ 6= e ensures that self-loops are represented by a pair of edges. Bidirected graphs have an even number of edges. An edge $e = (v, w)$ is said to have *source* v and *target* w and it is said to be an edge *out of* v and *in to* w. A pair $\{e, rev(e)\}$

represents an edge of the underlying undirected graph and we therefore call such a pair a *uedge*.

A *map* is a bidirected graph together with a cyclic ordering on the edges out of v for every vertex v.

Figure. 2 shows a map.

For a map G we define two successor operations on the edges of G. Let e be an edge with source v. The *cyclic adjacency successor* $as(e)$ and *predecessor* $ap(e)$ of e are the edges following and preceding e in the cyclic ordering of the edges out of v. The *face cycle successor* $fs(e)$ is defined as $ap(rev(e))$ and the *face cycle predecessor* $fp(e)$ is defined as $rev(as(e))$. Then $as(ap(e)) = e$ and $fs(fp(e)) = e$.

The *face cycle* $fc(e)$ of an edge e is the shortest sequence $e0, e1, \ldots, ek{-}1$ of edges such that $e0 = e$, $ei{+}1 = fs(ei)$ for $0\ 6\ i < k - 1$ and $fs(ek{-}1) = e0$. The face cycles partition the edges of a graph.

For a map M let n be the number of nodes, nz be the number of nodes of degree 0, m be the number of edges, f be the number of face cycles, and c be the number of connected components. Then, $g = (m/2 - n - nz - f + 2c)/2$ is called the *genus*[2] of M, see [19].

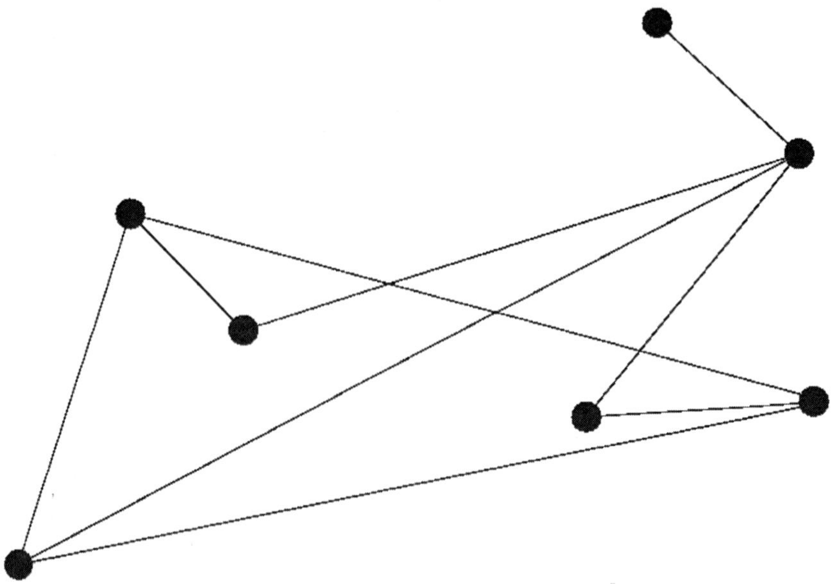

Figure 3. A locally consistent embedding that is not a planar embedding.

A geometric embedding (or simply embedding) of a bidirected graph assigns a point to each vertex and a Jordan curve connecting its endpoints to each uedge of the graph. An embedding is called a planar embedding if distinct points are assigned to distinct vertices and if the Jordan curves assigned to the edges are pairwise disjoint except for common endpoints. A planar embedding of a map M must satisfy the additional constraint that for each vertex v of M the cyclic ordering of the edges out of v agrees with the counter-clockwise ordering of the associated Jordan curves around v. An embedding is called *straight- line* if each edge is realized as a straight-line segment. It is well known that a map can be embedded into the plane if and only if its genus is zero.

Consider a planar embedding of a map M. The planar embedding partitions the plane into regions (=the connected components of R^2 minus the image of the embedding), one for each face cycle of M. The region corresponding to a particular face cycle lies locally to the left of the face cycle as one traverses the face cycle, cf. **Figure**. 2.

We derive two algorithms. Let M be a planar map. First, the genus of M can clearly be computed in linear time. Second, let *pos* assign a point to each vertex of M; then *pos* induces a straight-line embedding of M. We call *pos* a *locally consistent embedding* of M if the two endpoints of each edge are assigned to distinct points in the plane and if for each vertex v of M the counter-clockwise ordering of the neighbors of v around v agrees with the cyclic ordering of the edges out of v. In order to verify that *pos* is a locally consistent embedding of M it suffices to check that

(1) for each edge $e=(v, w)$ of M distinct points are assigned to v and w, and that

(2) for each vertex v of M the sequence of directions defined by the edges out of v is counter-clockwise ordered.

Both conditions are easily checked in linear time. We close this section with a warning. A locally consistent embedding is not necessarily a planar embedding as **Figure**. 3 shows.

Convex planar subdivisions, Delaunay triangulations, anDelaunay diagrams

Let $M=(V, E)$ be a connected map and let $pos: V \rightarrow R^2$ define a straight-line planar embedding of M. The pair (M, pos) is called a *convex planar subdivision* if each bounded region of the embedding is a convex polygon and if the complement of the unbounded region is a weakly convex polygon, see **Figure**. 4.

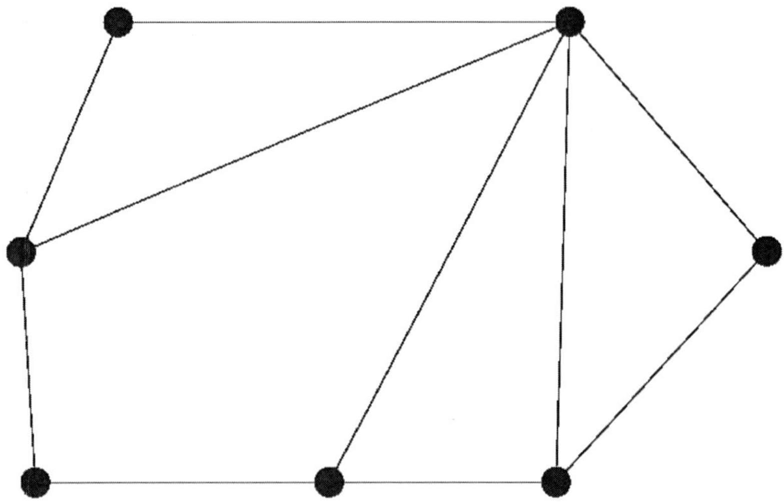

Figure 4. A convex subdivision.

The embedding is called a *triangulation* if, in addition, each bounded region is a triangle. It is called a *Delaunay triangulation* if it is a triangulation and if the circumcircle of each triangle contains no other vertex in its interior. It is called a *Delaunay diagram* if it is a convex planar subdivision such that every completion of *M* (= a completion of *M* is a super-graph of *M* in which each bounded region is a triangle) is a Delaunay triangulation and such that no subgraph of *M* has this property. We refer the reader to [7, 15] for further information about Delaunay triangulations anDiagrams. We derive algorithms that check whether a pair *(M, pos)* is a planar subdivision, a triangulation, a Delaunay triangulation, and a Delaunay diagram, respectively.

Convex planar subdivisions and triangulations. We first derive a necessary condition for convex planar subdivisions. Assume that *(M, pos)* is a convex planar subdivision. Consider any face cycle *fc* of *M*. If the associated region is bounded then *fc* is a counter-clockwise oriented convex polygon and if the associated region is unbounded then *fc* is a clockwise oriented weakly convex polygon.

Lemma 1. *Let M be a map and let pos assign points in the plane to the nodes of M. Then (M, pos) is a convex planar subdivision if the following conditions are satisfi*

- *M is connected and has genus zero.*
- *pos is a locally consistent embedding of M.*
- *Each face cycle is either a counter-clockwise convex polygon or a clockwise weakly convex polygon. There is exactly one face cycle of the latter kind.*

Proof. Let R be the open region that is enclosed by the unique face cycle *fc* that is a weakly convex clockwise polygon. [3] We claim that all vertices of M that are not a vertex of *fc* lie in R. Assume otherwise. Then there must be a vertex v that is not part of *fc* and a direction d such that v is a maximal vertex of G in direction d (note that we said "a maximal vertex" and not "the maximal vertex"). Since v is maximal there must be a pair of edges incident to v which span an angle of at least π and hence v must be part of a weakly convex chain. Since *fc* is the unique weakly convex chain v belongs to *fc*, a contradiction.

Now we distinguish cases. First assume that R is non-empty. Each face cycle of G different from *fc* defines a counterclockwise oriented convex polygonal region in the plane. We need to show that these regions form a partition of R. Consider a point p moving in the plane such that it avoids vertices of G. Whenever p crosses a directed edge e it will enter another region (namely, the one to the left of *rev(e)*) except when *rev(e)* belongs to *fc*. This shows that all points in the interior of R are covered by the same number of regions. Also, since all vertices on the boundary of R are part of *fc*, exactly one bounded region is incident to each edge of *fc*. Altogether we have shown that the regions defined by the face cycles different from *f* partition R and hence *(M, pos)* is a convex planar subdivision.

Next assume that R is empty. Since *fc* is a weakly convex polygon this implies that all vertices of *fc* lie on a common line and that for all but two edges of *fc* the face cycle successor of the edge points into the same direction as the edge itself. Thus *(M, pos)* is a convex planar subdivision.

All conditions of Theorem 1 are readily checked in linear time. For a triangulation, we need to check in addition that every counter-clockwise convex face cycle is a triangle. We therefore have the following theorem.

Theorem 1. *It can be checked in linear time whether a pair (M, pos) defi a convex planar subdivision or a triangulation.*

Delaunay triangulations anDelaunay diagrams. Let *(M, pos)* be a triangulation. It is a Delaunay triangulation if for any two triangles sharing an edge the third vertex of the first triangle is not contained in the interior of the circumcircle of the other triangle. This is easily checked in linear time.

For Delaunay diagrams we have to work slightly harder. We review some facts about Delaunay diagrams. Let S be a set of points in the plane and let s and t be two points in S. The segment st belongs to some Delaunay triangulation of S if there is a circle passing through s and t such that the interior of the circle contains no point of S and the segment belongs to every Delaunay triangulation of S if there is a circle passing through s and t such that the closure of the circle contains no points of S other than s and t. The Delaunay diagram of S consists of all segments st that belong to every Delaunay triangulation of S. In particular, the Delaunay diagram is unique.

Lemma 2. *A convex planar subdivision M is a Delaunay diagram if and only if*

- *the vertices of each bounded face are co-circular, and*
- *for every pair of bounded faces f and g sharing an edge e the endpoint of the face cycle successor of e on g lies outside the circle passing through the vertices of f.*

Proof. The conditions are clearly necessary. We show sufficiency.

Consider any completion C of M. It is obtained by triangulating the bounded faces of M. Let e be any edge of the completion and assume that e is shared by two triangles $T1$ and $T2$. If e was not an edge of M then $T1$ and $T2$ have the same circumcircle. If e was already an edge of M then let f and g be the bounded faces of M incident to e. The vertices of f and g are co-circular and, by the second condition, one vertex of g lies outside the circumcircle of f. Thus all vertices of g lie outside the circumcircle of f. This implies that the third vertex of $T1$ lies outside the circumcircle of $T2$. We conclude that C is a Delaunay triangulation.

It remains to show that all edges of M are required for a Delaunay diagram. This follows from the observation that for each edge e of M there is a circle passing through the endpoints of e and such that only the endpoints of e are contained in the closure of the circle.

Theorem 2. It can be checked in linear time whether a pair (M, pos) defi a Delaunay triangulation or a Delaunay diagram.

Voronoi diagrams

We show how to check whether a map is a Voronoi diagram of a set of point sites, see [9,15,17,18] for information about Voronoi diagrams. We start by fixing a particular representation of Voronoi diagrams. A Voronoi diagram is represented as a planar map as follows: there is a vertex for each vertex of the Voronoi diagram and, in addition, a vertex "at infinity" for each ray of the Voronoi diagram. Vertices at infinity have degree one. The edges of the graph correspond to the edges of the Voronoi diagram. Each edge is labeled with the site owning the region to its left and each vertex is labeled with a triple of points.[4] For a "finite" vertex the three points are any three sites associated with regions incident to the vertex (and hence the center of the circle passing through the points is the position of the vertex in the plane) and for a vertex at infinity the three points are collinear and the first point and the third point of the triple are the sites whose regions are incident to the vertex at infinity. Let a and c be the first and third point of the triple respectively; a and c encode the geometric position of the vertex at infinity as follows: the vertex lies on the perpendicular bisector of a and c and to the left of the segment ac.

A graph G representing a Voronoi diagram must clearly satisfy the following conditions.

- G is a planar map which is locally consistent with the positions assigned to its vertices.
- The site information associated with edges is consistent, i.e., if e and e' are consecutive edges on some face cycle then both edges have the same associated site.
- The sites associated with e and *rev(e)* are distinct.
- Call a vertex whose associated triple of points is not collinear non-trivial and call it trivial otherwise. Every non-trivial vertex has degree at least three and every trivial vertex has degree one.
- For each non-trivial vertex each of the three defining points of the associated circle is associated with one of the incident edges and the sites associated with all incident edges lie on the associated circle.
- Each trivial vertex has an associated triple of the form $(a, \_, c)$, where a and c are distinct. Let e be the unique outgoing edge. Then the site associated with the face to the left of e is c and the site associated with the face to the right of e is a.
- For every edge $e = (v, w)$ such that v and w are nontrivial, the centers of the circles associated with v and w are distinct. Let p and q be these centers and let a be the site associated with e. Then a lies to the left of the segment pq.

When a graph G satisfies all conditions above we can construct a geometric object from it as follows. We assign a position *pos(v)* to each nontrivial vertex v and a segment, ray, or line *geo(e)* to each edge e. For a nontrivial vertex v let *pos(v)* be the center of the circle associated with v. For an edge $e = (v, w)$ let a and c be the sites separated by e, i.e., one of a and c is associated with e and the other with *rev(e)*. If v is nontrivial then a and c lie on the circle associated with v and hence *pos(v)* lies on the perpendicular bisector of a and b. Define *geo(e)* as follows. First assume that v and w are both nontrivial. Then *geo(e)* is the segment directed from *pos(v)* to *pos(w)*. Note that this segment has nonzero length and is part of the perpendicular bisector of a and c. Next assume that exactly one of v and w is nontrivial. Assume without loss of generality that the triple of points associated with the trivial point is of the form *(a,_, c)*. If w is trivial then *geo(e)* is the ray starting at *pos(v)*, running along the perpendicular bisector of a and c, and extending to infinity to the left of the segment ac. If v is trivial then *geo(e)* is the ray ending in *pos(w)*, running along the perpendicular bisector of a and c, and coming from infinity to the left of the segment ac. Finally, assume that v and w are trivial and assume without loss of generality that the triple of points associated with v is of the form *(a,_, c)*. Then *geo(e)* is the bisector of a and c oriented such that a lies to its left.

Now we distinguish cases according to whether G is connected or not.

First assume that G is connected. Define a face chain as a minimal sequence $e0, e1, \ldots, ek$ of edges such that $ei{+}1$ is the face cycle successor of ei for all i, $0\ 6\ i < k$, and either $target(ek) = source(e0)$ or $source(e0)$ and $target(ek$ $)$ have degree one. We call face chains of the former kind closed and face chains of the latter kind open. We have the following additional necessary conditions: All face chains must be strictly convex counterclockwise oriented and the rays going to infinity must wind around the origin only

once. We claim that these conditions are also sufficient for G being a Voronoi diagram.

Consider any face chain f. All edges on the boundary of f have the same associated site, say a, the circles associated with all nontrivial vertices of f pass through a, for each edge e of f, *geo(e)* is part of the perpendicular bisector of a and the site associated with the other side of e, and a lies to the left of *geo(e)*. Define

$$reg(f) = \bigcap_{e;\ e \text{ is an edge of } f} H(a, site\_of\_rev(e)).$$

where *site_of_rev(e)* is the site associated with the reversal of *e* and *H (a, site_of_rev(e))* is the half- plane defined by *a* and *site_of_rev(e)* and containing *a*. Then *reg(f)* is a convex polygonal region which contains the Voronoi region of site *a* (since in the definition of Voronoi region the intersection is over all sites different from *a*). We still need to show that the regions partition the plane. Consider a point moving in the plane and avoiding vertices of regions. Such a point is always covered by the same number of regions. Moreover, when the point travels along a cycle at infinity it is always covered by exactly one region since the rays of the diagram wind around the origin once. Altogether we have shown that the regions partition the plane and hence *G* is a Voronoi diagram.

We turn to the case that *G* is not connected. If *G* is not connected it can only be the Voronoi diagram of a set of collinear sites. As such it must have the additional property that no site is associated with three edges of *G*, that exactly two sites are associated with exactly one edge of *G* (alternatively, that the number of distinct sites is equal to $m/2 + 1$), and that for each site *s* that is associated with two edges of *G* the sites associated with the reversals of these edges lie on opposite sides of *s*. We show that these conditions, i.e., all sites are collinear, no site is associated with three edges, exactly two sites are mentioned once, and for each site that is mentioned twice the neighbors lie on opposite sides of it, suffice.

Clearly, the geometric interpretation of *G* is a set of parallel lines. Consider the placement of the sites on their common underlying line. For each site *s* which is associated with two edges it is guaranteed that the two adjacent sites (= sites for which there is an edge having *s* on one of its sides) lie on opposite sides of *s*. Thus the sites are nicely ordered along their underlying line.

We summarize.

Theorem 3. *It can be checked in time* $O(n \log n)$ *whether a graph G with n nodes represents a Voronoi diagram. Linear time suffi if the graph is connected.*

Proof. We have already argued that the conditions introduced above are necessary and sufficient. If *G* is connected then all conditions can be checked by graph traversals and hence linear time suffices. When *G* is not connected we need to determine the number of distinct sites that are associated with the edges of *G*. The simplest way to accomplish this task is to use a dictionary indexed by sites. Thus each edge contributes time $O(\log n)$ to the time bound.

Some authors augment the representation of Voronoi diagrams with a cycle at infinity. This slightly more verbose representation is always connected and can be checked in linear time in all cases. We leave the details to the reader.

Line segment intersection

Trapezoidal decompositions are the output of line segment intersection algorithms. We show how to verify whether a given planar map represents a trapezoidal decomposition and whether it represents the trapezoidal decomposition for a given set of segments.

A *trapezoidal decomposition* T is a subdivision of the plane, in which all regions have vertical trapezoidal shape. A region has *vertical trapezoidal shape*, if it is a halfspace bounded by a vertical line, or a vertical strip, or the intersection of a vertical strip and a halfspace, or the intersection of a vertical strip and two halfspaces whose boundaries do not intersect in the interior of the strip.

A trapezoidal decomposition is represented as a planar map M as follows. There is a vertex for each vertex in the trapezoidal decomposition and, in addition, a vertex of degree one for each ray going to infinity. The geometric position of vertices at infinity is recorded as in the case of Voronoi diagrams. We show how to check whether a given planar map M represents a trapezoidal decomposition.

Let us assume first, that M is connected. M must satisfy the following conditions.

- M has genus zero and M is a locally consistent embedding.
- All face cycles but one form a trapezoid, i.e., they consist of a sequence of vertical edges with increasing y-coordinate in the embedding, followed by a sequence of collinear edges with decreasing
- x-coordinate, followed by a sequence of vertical edges with decreasing y-coordinate, followed by a sequence of collinear edges with increasing x-coordinate.
- All degree one vertices in M lie on the remaining face cycle. The degree one vertices split the face cycle into minimal subsequences of edges bounded by degree one vertices. There are exactly two such subsequences where all edges have a vertical embedding. In one of them, the edges have increasing y-coordinate, in the other one the y-coordinate is decreasing. Let v_{ds}(v_{dt}) be the vertex where the decreasing sequence of vertical edges starts (ends). Similarly, define v_{is} and v_{it}. All edge subsequences on the face cycle between v_{dt} and v_{is} have the following structure. They start with a sequence of vertical edges with increasing y-coordinate, followed by a sequence of collinear edges with decreasing x-coordinate, followed by a sequence of vertical edges with decreasing y-coordinate. Analogously, all edge subsequences on the face cycle between v_{it} and v_{ds} have the following structure. They start with a sequence of vertical edges with decreasing y-coordinate, followed by a sequence of collinear edges with increasing x-coordinate, followed by a sequence of vertical edges with increasing y-coordinate.

Lemma 3. *If M is connected and satisfi the conditions above, then M represents a trapezoidal decomposition.*

Proof. We want to apply Lemma 1. To this end we conceptually modify and extend *M*. Let y_{min} and y_{max} be the minimum and maximum *y*-coordinate assigned to any vertex of *M*, respectively. We modify the embedding of the unique face cycle containing all degree one vertices, by giving all degree one vertices between v_{it} and v_{ds} the *y*-coordinate $y_{max}+1$ and all degree one vertices between v_{dt} and v_{is} the oordinate $y_{min}-1$ (the *x*-coordinates of these vertices are not changed) and by connecting consecutive degree one vertices by a new edge. Let M' be the resulting embedding. In M' the face cycle containing all former degree one vertices defines a region whose complement is a weakly convex polygon. Furthermore, all other face cycles define a convex region.

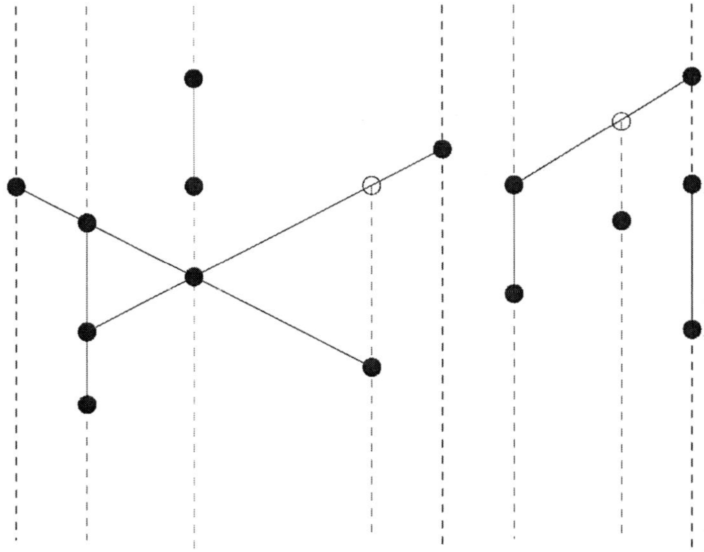

Figure 5. The trapezoidal decomposition induced by a set of segments. Primary vertices and edges are drawn solid and secondary vertices and edges are drawn as empty circles anDashed lines, respectively. The degree one vertices at infi ity are not shown.

Thus, by Lemma 1, M' is planar. Since the union of the new trapezoids in M' and the unbounded face is exactly the unbounded face of *M*, we are done.

If *M* is not connected, we apply the above procedure to each connected component. Then we check whether the *x*-ranges of the connected components are disjoint. If they are not, we reject the map, if yes, the map represents a trapezoidal decomposition.

We turn to trapezoidal decompositions induced by a set of line segments. Let S be a set of non- overlapping segments,[5] i.e., any two segments in S have at most one point in common. We allow segments of length zero. The trapezoidal decomposition $T(S)$ induced by S is defined as follows, see **Figure**. 5 for an illustration. Each vertex or edge of $T(S)$ is either *primary* or *secondary*. The primary vertices are the endpoints of the segments in S and the intersections between the segments in S. For each primary vertex v that has upward (downward, respectively) sight, i.e., for which there is no vertical segment in S containing v and not having v as its upper (lower) endpoint, there is a secondary edge e incident to v. This edge is vertical, has v as its lower (upper) endpoint, and ends at the next segment above (below) v, if any, or extends to infinity. If e ends at some segment in S and the other endpoint of e is not already a primary vertex then it is a secondary vertex in $T(S)$. The primary edges of $T(S)$ are the maximal subsegments of segments in S containing no (primary or secondary) vertex of $T(S)$ in their interior. This completes the definition of $T(S)$.

In the representation of $T(S)$ by a planar map we label each edge as primary or secondary and we label each primary edge by the segment containing it. We also label each primary vertex that corresponds to a segment of length zero.

A map M representing a trapezoidal decomposition $T(S)$ must satisfy the following additional conditions.

- The segment labeling is consistent, i.e., for each edge e labeled with a segment we have $seg(e) = seg(rev(e))$.
- Edges labeled as secondary have a vertical embedding.
- If e is labeled with segment s then the embedding of e is supported by s.
- If a vertex v is labeled with a segments s, then s is trivial and the position assigned to v is coincident with s.
- The vertices fall into three classes A, B and C (which correspond to the vertices at infinity, the primary vertices, and the secondary vertices, respectively). Class A contains all degree one vertices. Class B contains all which either have incident edges labeled with at least two different segments or have only one incident edge labeled with a segment or are labeled with a segment. Class C contains all vertices having exactly two incident edges labeled with a segment and where moreover both segment labels are the same.
- Class B vertices have two incident vertical edges, class C vertices have at least one.
- Every vertical edge has at least one endpoint in class B.
- For each segment $s \in S$ the union of the edges labeled by s is equal to s.

Theorem 4. *It can be checked in time* $O(n \log n + k)$ *whether a map M represents the trapezoidal decomposition T (S) of a set S of segments. Here n denotes the number of segments and k denotes the number of intersections.*

Proof. Assume that M satisfies all conditions set forth above. Then M represents a trapezoidal decomposition, the union of the primary edges is equal to the union of the segments in S, and each secondary edge is justified by some primary node. Assume that some intersection of two segments is not present in M. Then the two segments are either modeled by edges in different connected components of the graph or are modeled by edges in a single connected component. In the first case we would have overlapping x-ranges and in the second case we would have a contradiction to the fact that M is an embedded planar map.

It remains to argue the time bound. All but the disjointness of x-ranges and the fact that each segment in S is fully represented in the map is easily checked in linear time. Disjointness is easily checked in time $O(n \log n)$ by sorting. The condition that each segment is fully represented in the map can be checked in linear time as follows. For each vertex v we run over all incident edges and link edges that are labeled by the same segment. This can be done in time proportional to the degree of the edge. Having linked all the

edges labeled by the same segment it is easy to traverse all edges supported by the same segment.

Some authors define $T(S)$ as being enclosed in a large box. This makes $T(S)$ connected and reduces the checking time to linear.

2.3. Verifi of higher-dimensional convex polyhedra and convex hulls

We first deal with the question whether a simplicial piecewise linear hyper-surface F without boundary in d-dimensional space is the boundary of a convex polytope. We assume that the hyper-surface is given by its facet graph. The facet graph is a d-regular graph whose nodes are labeled by d-tuples of affinely

independent points, i.e., each node corresponds to an oriented $(d-1)$-simplex (= a facet of the surface). The hyperplane supporting a facet divides d-space into a positive and a negative halfspace. Neighboring nodes differ in their labeling by exactly one point and for everyone of the d vertices of a facet there must be such a neighboring facet. In other words, edges correspond to $(d-2)$-simplices shared by two facets.

Neighboring nodes must be labeled consistently, i.e., the negative halfspace corresponding to adjacent facets must agree locally.

Task 1. *Given the facet graph of a simplicial piecewise linear hyper-surface* F *in d-dimensional space decide whether* F *is the surface of a convex polytope.*

We take a theorem of differential geometry as the tarting point of our solution.

Theorem 5 (Hadamard). *Let* F *be a smooth compact surface in* R^d *without boundary and let* $d > 2$. *If* F *is locally convex at everyone of its points then* F *is the surface of a convex body.*

This theorem suggests that it suffices to check local convexity at every ridge of a simplicial surface. Although this is clearly a necessary condition for global convexity it is not sufficient as **Figure**. 6 shows, i.e., the non-smoothness of simplicial surfaces complicates matters. We will prove the following theorem for the polyhedral case.

Theorem 6. *Let* F *be a simplicial* $(d - 1)$*-dimensional surface without boundary in* R^d *that is consistently oriented, let o be center of gravity of all the vertices of surface* F *and let p be the center of gravity of some facet of* F. *Then* F *is the surface of a convex body iff*

- F *is locally convex at all its ridges,*
- *o is on the negative side of all its facets, and*
- *the ray emanating from o and passing through p intersects only the facet containing p.*

Before we prove this result we illustrate it on **Figure**. 6. Let *o* be any point in the *x, y*-plane that is on the negative side of every facet of the surface shown. All but two rays emanating from *o* intersect the surface twice and hence witness the non-convexity of the surface. The two exceptional rays go through the two tips of the bipyramid, i.e., pass through a lower dimensional feature of the surface. The key insight underlying our criterion is that this observation is generally true.

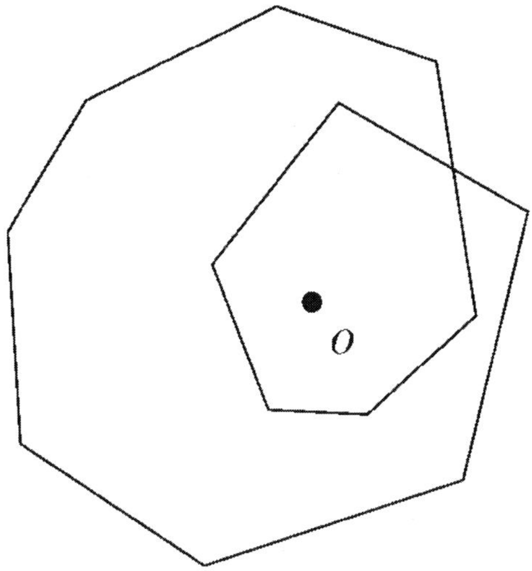

Figure 6. Local convexity at ridges does not suffice: The **Figure**ure shows a locally convex yet self-intersecting polygon in the x, y-plane. A smooth version of this curve demonstrates that the condition $d > 2$ is necessary in Theorem 5. Extending the polygon to a bipyramid by adding two points, one each on the negative and positive z-axis, and constructing two triangles for each polygon edge yields a simplicial surface that is locally convex at every ridge but has self-intersections. Thc point o shown is on the negative side of all facets of the surface.

The conditions listed in Theorem 6 are clearly necessary. We will prove their necessity as follows. We first show that the last two conditions imply that F is the boundary of a star-shapeD-polytope. Next we show that a star-shaped polytope that is convex at all its ridges is convex.

Lemma 4. *Let* F *be a simplicial piecewise linear hyper-surface without boundary in* R^d *that is consistently oriented, and let o be a point that is on the negative side of each facet of* F.

Every ray emanating from o that intersects facets of F *only in their relative interiors intersects* the same number *of facets of* F.

Proof. For each facet f of F let cf be the cone with apex o spanned by f. Let $A(o, F)$ be the arrangement formed by all those cones. Each d-dimensional cell c in $A(o, F)$ is a cone with apex o. By construction every ray emanating from o that lies in c intersects the same number of facets of F, say n_c (although the order in which a ray intersects these facets is not constant over c).

We now claim that $n_c = n_{c'}$ for all cells c, c' in $A(o, F)$. It suffices to show this for cones c, c' that share a $(d-1)$-face, say g. In the generic case g must be part of a cone with apex o that is spanned by some ridge of F. This ridge is contained in exactly two facets f, f' of F, one of which, say f, intersects c, and the other intersects c'. (This is a consequence of F being consistently oriented and of f and f' both having o on the same side.) Since except for f and f' the remaining facets of F that intersect c are the same as the ones that intersect c' we can conclude that in the generic case we indeed have $n_c = n_{c'}$. In the non-generic case g can be part of several cones spanned by ridges of the surface F. But as before for each of those ridges one of the two containing facets intersects c and the other intersects c' with the remaining facets intersecting c being the same as the ones for c'.

Corollary 7. *If the last two conditions of Theorem* 6 *hold then* F *is the boundary of a star-shaped polytope whose kernel contains o.*

In view of the above it suffices to prove the following. Let F be a simplicial $(d-1)$-dimensional surface without boundary in $\mathbb{R}^d$ that is consistently oriented, and let o be a point that is on the negative side of each facet of F. Assume further that F is locally convex at all its ridges and that every ray emanating from o that intersects facets of F only in their interiors intersects exactly one facet of F. Then F is the surface of a convex body.

Lemma 5. *For every facet f of* F *let h_f be the supporting hyperplane, let A be the arrangement defi by these hyperplanes, and let K be the cell of A containing o. Then every facet of K is part of a facet of* F.

Proof. Assume otherwise. Let g be a facet of K that is not contained in any facet of F and let C be the other cell of A incident to g. There must be a facet f of F such that g is contained in h_f and a ridge r of f that separates f from g in h_f. Let f' be the other facet of F incident to r and view the situation in the 2-plane orthogonal to r. In this plane r is a point and f and f' are edges incident to r, K is in the negative halfspace of f and f', g is a line segment on the boundary of K, on the same line as f and separated from f by r. Thus F is not locally convex at r.

We next strengthen the preceding lemma.

Lemma 6. *The boundary of K is contained in* F.

Proof. Let x be any point in the boundary of K. Then x is contained in the closure of some facet of K

and hence in the closure of some facet of F.

We complete the argument by showing the following lemma.

Lemma 7. F *is the boundary of K.*

Proof. The boundary of K is contained in F. Also F and the boundary of K are closed simplicial surfaces. Hence they are equal.

The verification procedure described above is easy to program.

- Check local convexity at every ridge. If local convexity does not hold at some ridge declare F non- convex.
- Set o to the center of gravity of the vertices of F and check whether o is on the negative side of all facets. If not, declare F non-convex.
- Choose any facet and let p be the center of gravity of its vertices. Let r be the ray emanating from o
- and passing through p. If r intersects the closure of any other facet of F declare F non-convex.
- If F passes both tests declare it the surface of a convex polytope. We next turn to the convex hull problem.

Task 2. *Given a set S of points and a hyper-surface* F *verify that* F *is the boundary of the convex hull of S.*

We know already how to check the convexity of F. So let us assume that F has passed the convexity test and let P be the convex polytope whose boundary is F. It suffices to verify that

- every vertex of P is a point in S, and that
- every point of S is contained in P.

The first item is fairly easy to check. If the vertices of P are specified as pointers to elements in S the check is trivial. If the vertices in S are specified by their coordinate tuples the check involves a dictionary lookup. This is a first example where the representation of the output influences the complexity of testing. The second condition is much harder to check. In fact, we found no easy and efficient method to verify it. A simple method would be to check every point of S against every facet of F. However, the complexity of this method is an order of magnitude larger that the complexity of the most efficient convex hull programs.[6] An alternative method is to use linear programming to check that all non-vertices are non-extreme.[7] For fixeDimension the alternative method is quadratic in the number of vertices. For variable dimension one might hope that a simplex-based verification procedure has good expected running time. Nevertheless, both approaches essentially resolve the original problem. We conclude that convex hull programs that output the hull as a convex polytope are hard to check. Gift-wrapping falls in this category. Can a convex hull program help the checker? The algorithms based on randomized incremental construction compute a simplicial complex[8] comprising the hull, i.e., a set of simplices whose union is P. They also compute for each point of S a simplex of the complex containing the point. With this representation of the hull the second item is easy to check. One only has to check whether every point of S is indeed contained in the simplex it is supposed to be in. This amounts to the solution of a linear system.

LAYERS OF SOFTWARE AND EXPERIMENTAL RESULTS

We implemented most of the program checkers described in this paper. The implementations can be found in [10,12]. We report on our experiences with the checker for convex hulls. The hull program [10] is structured into four layers: arbitrary precision integer arithmetic, linear algebra over the rationals, geometric primitives, and the convex hull algorithm proper. We added program checking to the three top layers.

The linear algebra package realizes the basic functions of linear algebra (solution of homogeneous and non-homogeneous systems, matrix inversion, determinant of a matrix, rank of a matrix) for matrices with integer entries. Many checks are trivial,[9] e.g., it is trivial to check whether $x0$ solves $Ax = b$. More interesting are negative answers. In order to allow verification of the statement that the system

$Ax = b$ has no solution the solver provides a vector c such that $c^T A = 0$ and $c^T b$ 6= 0. Such a vector c is guaranteed to exist and is readily computeDuring Gaussian elimination. [10] In order to allow verification of a determinant the determinant program also returns an LU-decomposition of the matrix.

Among the geometric primitives we have the orientation test for $d + 1$ points in d-space, the computation of the hyperplane passing through a set of d-points, and the membership test for simplices.

Given the checking abilities of the linear algebra layer the geometric primitives are easy to check.

The hull program realizes the algorithm due to [4,5]. It is checked according to the theory developed in the preceding section.

We implemented the checkers for the linear algebra and geometric primitives layer concurrently with the development of these layers. We found that debugging time was significantly decreased by the improved localization of errors that the checkers provided. The high level checker was implemented after the program was completed anDebugged visually for 2- and 3-dimensional inputs. In fact, our investigation was initiated by the question whether successful testing in low dimensions has any implications for higher dimensions. Checking at all levels increases the running time of the hull program by a factor of less than two.

Large software systems are built in layers. As the number of layers increases debugging becomes harder and harder as errors become harder and harder to locate. If all programs in a system are self- checking any program can safely assume that all the lower level programs used by it are correct. *We believe that program checking will play an important role in the construction of software libraries and have started to add checking facilities to LEDA. We also believe that programs in a library should be designed for check-ability.*

Can we make all programs in a library self-checking? There are two obstacles to that. Firstly, there might be problems where verification brings about an unbearable overhead. Secondly, programs in a library might have to rely on user provided functions over which the library has no control. The data structure area provides examples for this situation as we will see in the next section.

DICTIONARIES AND COMPARE FUNCTIONS

Consider implementations of the dictionary data type. A dictionary realizes a finite subset S of some key type K under the operations insert, delete, and member. For a key x *member(x)* returns true iff $x \in S$, *insert(x)* adds x to S, anD*elete(x)* removes x from S. Comparison-based implementations of the dictionary data type require the key type K to be linearly ordered. To this end the user of the dictionary type provides a function *compare* that is assumed to define a linear order on K, i.e., *compare(x, x)* $= 0$ for any key x, *compare(x, y)* $\in \{-1, +1\}$ and *compare(x, y)* $=$ *−compare(y, x)* for any distinct keys x and y, and *compare(x, y)* $=$ *compare(y, z)* $= -1$ implies *compare(x, z)* $= -1$ for all keys x, y, and z. The function *compare* is user defined and therefore outside the control of the dictionary class.

Task 3. *Provide a dictionary implementation that is immune against errors in the compare function, i.e., it either performs all dictionary operations correctly or reports an error in the compare function. The implementation may assume that compare is a function, i.e., all calls compare(x, y) return the same value.*

A first approach is to always make all pairwise comparisons between all elements in the current set and to abort as soon as the compare turns out to be incorrect. This approach has a linear overhead for insertions. Can we do better?

Lemma 8. *One can guard against errors in the compare function and still have logarithmic search and update time.*

Proof. At any time we store the current set S in the nodes of a binary tree of logarithmic depth. We maintain the invariant that for any pair of vertices of the tree such that one is an ancestor of the other the comparison between these keys was explicitly made and had the expected answer. Thus $O(n \log n)$ out of the n^2 comparisons are explicitly known. It follows immediately from the invariant that searches for elements in the current set S are successful and searches for elements not in S end in a leaf where the element can be put without destroying the invariant. Thus member queries and insertions (without rebalancing) work correctly.

Assume now that our class of trees is rebalanced by rotations. When we rotate at a node v we need to evaluate the compare function for all pairs of nodes that become related (in the ancestor descendant relationship) by the rotation. If any of these evaluations has an unexpected result, e.g., an element that is a left descendant of some other element in the tree is declared larger by *compare*, a witness to the non- transitivity of the compare function is reported and execution is aborted. With this scheme the cost of a rotation at a node v is proportional to the size of the subtree rooted at v. Even with this enormous cost of rotation the amortized cost of an insertion is still logarithmic (at least for weight-balanced trees, randomized search trees and skiplists). For weight-balanced trees this was shown in [8, p. 198, Theorem 5], for randomized search trees this was shown in [1], and for skiplists this was shown in [13].

Deletions are also easy to handle: rotate the element to be deleted to the boundary of the tree and delete. Again the amortized cost of a deletion stays logarithmic.

Finger trees are a powerful extension of standard search trees, see any of the references above. Finger trees support sublogarithmic searches and updates. The performance of finger trees relies heavily on the fact that any of the three data structures mentioned above support updates at a specified position of the tree in constant amortized or expected time (provided the cost of a rotation is constant).

We now argue that the efficiency of finger trees cannot be maintained if we want to safeguard against errors in the compare function.[11] Note first that any tree baseData structure must make all the comparisons between related nodes (if one misses one pair an adversary could force a failed lookup that should have been successful). Since there are • *(n* log *n)* such pairs, *n* insertions into an initially empty tree must take • *(n* log *n)* time. Thus the efficiency of finger trees can not be maintained.

CONCLUSION

We described efficient and simple program checkers for some of the basic problems of computational geometry. We have seen that for some problems the output representation has a considerable influence on the complexity of the verification task. Thus *algorithms should be designed with the verifi*

We have incorporated a checking facility into three of the four software layers of a convex hull program. Addition of checking reduceDebugging time and increased our confidence in the correctness of the implementation. We argued that *program checking in a library environment where software is developed in layers is of utmost importance.*

We showed that comparison baseDictionary implementations can guard against errors in the user- provided compare function and still keep logarithmic search and update time. We also argued that the sublogarithmic update time of finger searches cannot be maintained.

NOTES

[1] Formal verification of simple checkers might be a realistic goal in the near future.

[2] Recall that a map is a directed graph. The underlying undirected graph has *m/*2 edges. Note also that each vertex of degree zero can be removed by adding two edges. This increases *m/*2 by one, decreases *nz* by either one or two depending on whether

the vertex is connected to a vertex of degree larger than zero or not, leaves f unchanged or increases it by one (if nz decreases by two), anDecreases c by one. Thus, g does not change. When the number of connected components is one the formula for the genus reduces to the well-known Euler formula.

3 R may be empty.

4 It would suffi e to label points at infinity with a pair of points. For uniformity and for compatibility with LEDA we use a triple of points for every vertex.

5 We assume that overlapping segments have been partitioned into segments with disjoint interiors.

6 Algorithms based on randomized incremental construction [4–6] run in time related to the size of the output and the size of intermediate hulls and the algorithm of [16] is guaranteed to construct the hull in logarithmic time per face.

7 The linear program has d variables corresponding to the coefficients of a linear function. For each vertex of F there is a constraint stating that the function value at the vertex is negative. For each non-vertex consider the linear program that maximizes the function value at this point.

8 A simplicial complex is a set of simplices the intersection of any two is a face of both.

9 We assume that matrix–vector and matrix–matrix product are easy programs that need not be checked.

10 Linear system solving is probably the most striking example where a minor change in the output convention turns a program which is very difficult to check (after all, what do you know when a linear system is declared unsolvable) into a program that is trivial to check.

11 We refrain of calling this observation a lemma since we have not formally defined what we mean by a comparison based solution of the dictionary problem.

REFERENCES

1. C. Aragon, R. Seidel, Randomized search trees, in: Proc. 30th Ann. IEEE Symp. on Foundations of Computer Science, 1989, pp. 540–545.

2. M. Blum, S. Kannan, Programs that check their work, in: Proc. of the 21th Annual ACM Symp. on Theory of Computing, 1989.

3. M. Blum, M. Luby, R. Rubinfeld, Self-testing/correcting with applications to numerical problems, in: Proc. 22nd Annual ACM Symp. on Theory of Computing, 1990, pp. 73–83.

4. Ch. Burnikel, K. Mehlhorn, S. Schirra, On degeneracy in geometric computations, in: Proc. SODA '94, 1994, pp. 16–23.

5. K.L. Clarkson, K. Mehlhorn, R. Seidel, Four results on randomized incremental constructions, Computational Geometry 3 (1993) 185–212.

6. K.L. Clarkson, P.W. Shor, Applications of random sampling in computational geometry, II, J. Discrete Comput. Geom. (1989) 387–421.

7. B. Delaunay, Izv. Akad. Nauk SSSR 7 (1934) 793–800.

8. K. Mehlhorn, Data Structures and Algorithms 1: Sorting and Searching, Springer, Berlin, 1984.

9. K. Mehlhorn, Data Structures and Algorithms 3: Multidimensional Searching and Computational Geometry, Springer, Berlin, 1984.

10. K. Mehlhorn, M. Müller, S. Näher, J. Ziegler, An implementation of a convex hull algorithm, 1997 http://www.mpi-sb.mpg.de/~mehlhorn/Programs.html.

11. K. Mehlhorn, S. Näher, LEDA: A platform for combinatorial and geometric computing, Comm. ACM 38 (1) (1995) 96–102.

12. K. Mehlhorn, S. Näher, Checking geometric structures. http://www.mpi-sb.mpg.de/~mehlhorn/ Programs.html.

13. K. Mehlhorn, S. Näher, Algorithm design and software libraries: Recent developments in the LEDA project, in: Algorithms, Software, Architectures, Information Processing '92, Vol. 1, Elsevier, North-Holland, Amsterdam, 1992, pp. 493–505.

14. K. Mehlhorn, S. Näher, Ch. Uhrig, The LEDA user manual (version R 3.4). Technical Report, Max-Planck- Institut für Informatik, 1996. http://www.mpi-sb.mpg.de/LEDA/leda.html.

15. F.P. Preparata, M.I. Shamos, Computational Geometry: An Introduction, Springer, New York, 1985.

16. R. Seidel, Constructing higher-dimensional convex hulls at logarithmic cost per face, in: Proc. 18th Annu. ACM Sympos. Theory Comput., 1986, pp. 404–413.

17. G. Voronoi, Nouvelles application des parametres continus à la theorie the formed quadratic. premiere memoire: Sur quelqes proprietes des formes quadratique positives parfaites, J. Reine Angew. Math. 134 (1908) 198–287.

18. G. Voronoi, Nouvelles application des parametres continus à la theorie the formed quadratic. deuxième memoire: Recherches sur les parallélloèdres primitifs, J. Reine Angew. Math. 136 (1909) 67–181.

19. A.T. White, Graphs, Groups, and Surfaces, North-Holland, Amsterdam, 1973.

CHAPTER 2

Triangle-Free Geometric Intersection Graphs with No Large Independent Sets

Bartosz Walczak

Theoretical Computer Science Department, Faculty of Mathematics and Computer Science, Jagiellonian University

ABSTRACT

It is proved that there are triangle-free intersection graphs of line segments in the plane with arbitrarily small ratio between the maximum size of an independent set and the total number of vertices.

Keywords: Intersection graph Line segments Triangle-free Maximum independent set Fractional chromatic number

1 INTRODUCTION

Pawlik et al. [7] proved that there are triangle-free intersection graphs of line segments in the plane with arbitrarily large chromatic number. The graphs they construct have independent sets containing more than $1/3$ 1/3 of all the vertices. It has been left open whether there is a constant $c>0$ c>0 such that every triangle-free intersection graph of n n segments in the plane has an independent set of size at least cn cn. Fox and Pach [3] conjectured a much more general statement, that K_k Kk -free intersection graphs of curves in the plane have linear-size independent sets, for every k k. This would imply a well-known conjecture that k k -quasi-planar graphs (graphs drawn in the plane so that no k k edges cross each other) have linearly many edges [5], which is proved up to $k=4$ k=4 [1].

In this note, I resolve the independent set problem in the negative, proving the following strengthening of the result of Pawlik et al.:

Theorem There are triangle-free segment intersection graphs with arbitrarily small ratio between the maximum size of an independent set and the total number of vertices.

The constructions presented in the next two sections give rise to triangle-free intersection graphs of n n segments in the plane with maximum independent set size $\Theta(n/\log\log n)$ Θ(n/log log n).

2 CONSTRUCTION

Pawlik et al. [7] construct, for $k\geqslant 1$ k⩾1, a triangle-free graph G_k Gk and a family $\mathcal{P}_k$ Pk of subsets of $V(G_k)$ V(Gk), called *probes*, with the following properties:

1. $|\mathcal{P}_k| = 2^{2^{k-1}-1}$,

2. every member of $\mathcal{P}_k$ is an independent set of G_k,

3. for every proper coloring of the vertices of G_k, there is a probe $P\in\mathcal{P}_k$ such that at least k colors are used on the vertices in P.

They are built by induction on k, as follows. The graph G_1 has just one vertex v, and $\mathcal{P}_1$ has just one probe {v}. For $k\geqslant 2$, first, a copy (G,P) of $(G_{k-1},\mathcal{P}_{k-1})$ is taken. Then, for every probe $P\in\mathcal{P}$, another copy $(G_P,\mathcal{P}_P)$ of $(G_{k-1},\mathcal{P}_{k-1})$ is taken. There are no edges between vertices from different copies. Finally, for every probe $P\in\mathcal{P}$ and every probe $Q\in\mathcal{P}_P$, a new vertex d_Q connected to all vertices in Q, called the *diagonal* of Q, is added. The resulting graph is G_k. The family of probes $\mathcal{P}_k$ Pk is defined by

$$\mathcal{P}_k = \{P\cup Q\colon P\in\mathcal{P} \text{ and } Q\in\mathcal{P}_P\}\cup\{P\cup\{d_Q\}\colon P\in\mathcal{P} \text{ and } Q\in\mathcal{P}_P\}.$$

It is easy to check that the graph G_k is indeed triangle-free and the conditions (i)–(iii) are satisfied for $(G_k,\mathcal{P}_k)$—see [7] for details. It is also shown in [7] how the graph G_k is represented as a segment intersection graph.

I will show that there is an assignment w_k of positive integer weights to the vertices of G_k with the following properties:

1. the total weight of G_k is $\frac{k+1}{2}\cdot 2^{2^{k-1}-1}$,

2. for every independent set I of G_k, the number of probes $P \in \mathcal{P}_k$ such that $P \cap I \neq \emptyset$ is at least the weight of I I.

Once this is achieved, the proof of the theorem of this paper follows easily. Namely, it follows from (i) and (v) that every independent set I of G_k has weight at most $2^{2^{k-1}-1}$. We can take the representation of G_k as a segment intersection graph and replace every segment representing a vertex $v \in V(G_k)$ parallel segments lying very close to each other, so as to keep the property that any two segments representing vertices $u, v \in V(G_k)$ intersect if and only if $uv \in E(G_k)$ uv∈E(Gk). It follows from (iv) that the family of segments obtained this way has size $\frac{k+1}{2} \cdot 2^{2^{k-1}-1}$, while every independent set of its intersection graph has size at most $2^{2^{k-1}-1}$.

The assignment w_k of weights to the vertices of G_k is defined by induction on k, following the inductive construction of $(G_k, \mathcal{P}_k)$. The weight of the only vertex of G_1 is set to 1 1. This clearly satisfies (iv) and (v). For $k \geqslant 2$ k⩾2, let G, $\mathcal{P}$, G_P, P anD Q be defined as in the inductive step of the construction of $(G_k, \mathcal{P}_k)$. Let $p = |\mathcal{P}_{k-1}| = 2^{2^{k-2}-1}$. The weights w_k wk of the vertices of G are their original weights w_{k-1} in G_{k-1} multiplied by p p. The weights w_k wk of the vertices of every G_P GP are equal to their original weights w_{k-1} in G_{k-1}. The weight w_k wk of every diagonal d_Q dQ is set to 1 1. It remains to prove that (iv) and (v) are satisfied for $(G_k, \mathcal{P}_k, w_k)$ assuming that they hold for $(G_{k-1}, \mathcal{P}_{k-1}, w_{k-1})$.

The proof of (iv) is straightforward:

$$w_k(G_k) = w_k(G) + \sum_{P \in \mathcal{P}} (w_k(G_P) + |\mathcal{P}_P|) = 2pw_{k-1}(G_{k-1}) + p^2 = \tfrac{k+1}{2} \cdot 2^{2^{k-1}-1}.$$

For the proof of (v), let I be an independent set in G_k. Let $\mathcal{I} = \{P \in \mathcal{P} \colon P \cap I \neq \emptyset\}$. For every probe $P \in \mathcal{P}$, define

$$\mathcal{I}_P = \{Q \in \mathcal{P}_P \colon Q \cap I \neq \emptyset\}, \qquad \mathcal{P}'_P = \{P \cup Q \colon Q \in \mathcal{P}_P\} \cup \{P \cup \{d_Q\} \colon Q \in \mathcal{P}_P\},$$
$$D_P = \{d_Q \colon Q \in \mathcal{P}_P\}, \qquad \mathcal{I}'_P = \{P' \in \mathcal{P}'_P \colon P' \cap I \neq \emptyset\}.$$

By the induction hypothesis, we have

$$w_k(V(G)\cap I) \leqslant p|\mathcal{I}|, \qquad w_k(V(G_P)\cap I) \leqslant |\mathcal{I}_P|.$$

Suppose $P \in \mathcal{I}$. It follows that $(P\cup Q)\cap I \neq \emptyset$ and $(P\cup \{d_Q\})\cap I \neq \emptyset$ for every $Q \in \mathcal{P}_P$. Hence $|\mathcal{I}'_P| = |\mathcal{P}'_P| = 2p$. Moreover, we have $d_Q \notin I$ whenever $Q \in \mathcal{I}_P$, because d_Q is connected to all vertices in Q, one of which belongs to I. Hence

$$w_k(V(G_P)\cap I) + w_k(D_P\cap I) \leqslant |\mathcal{I}_P| + |\mathcal{P}_P \setminus \mathcal{I}_P| = |\mathcal{P}_P| = p.$$

Now, suppose $P \in \mathcal{P}\setminus\mathcal{I}$. If $Q \in \mathcal{I}_P$, then $(P\cup Q)\cap I \neq \emptyset$, $d_Q \notin I$ (by the same argument as above), and $(P\cup\{d_Q\})\cap I = \emptyset$. If $Q \in \mathcal{P}_P \setminus \mathcal{I}_P$, then $(P\cup Q)\cap I = \emptyset$, and $(P\cup\{d_Q\})\cap I \neq \emptyset$ if and only if $d_Q \in I$ dQ∈I. Hence

$$w_k(V(G_P)\cap I) + w_k(D_P\cap I) \leqslant |\mathcal{I}_P| + |D_P\cap I| = |\mathcal{I}'_P|.$$

To conclude, we gather all the inequalities and obtain

$$\begin{aligned} w_k(I) &= w_k(V(G)\cap I) + \sum_{P\in\mathcal{P}} \left(w_k(V(G_P)\cap I) + w_k(D_P\cap I)\right) \\ &\leqslant p|\mathcal{I}| + \sum_{P\in\mathcal{I}} p + \sum_{P\in\mathcal{P}\setminus\mathcal{I}} |\mathcal{I}'_P| = \sum_{P\in\mathcal{I}} |\mathcal{I}'_P| + \sum_{P\in\mathcal{P}\setminus\mathcal{I}} |\mathcal{I}'_P| = \sum_{P\in\mathcal{P}} |\mathcal{I}'_P|. \end{aligned}$$

3 IMPROVED CONSTRUCTION

Pawlik et al. [7] define also a graph $G\sim k$ G~k, which arises from $(G_k, \mathcal{P}_k)$ by adding, for every probe $P∈P\ k$, a diagonal $d\ P$ dP connected to all vertices in P. This is the smallest triangle-free segment intersection graph known to have chromatic number greater than k k. Define the assignment $\tilde{w}_k$ of weights to the vertices of $\tilde{G}_k$ so that $w\sim\ k$ w~k is equal to $w\ k$ wk on the vertices of $G\ k$ and $\tilde{w}_k(d_P) = 1$ for every $P∈P\ k$ P∈Pk. Let I I be an independent set in $\mathcal{I} = \{P \in \mathcal{P}_k \colon P\cap I \neq \emptyset\}$. Hence $d\ P \notin I$ dP∉I for $P∈I$ P∈I. It follows that

$$\tilde{w}_k(I) = w_k(V(G_k) \cap I) + |\{d_P \colon P \in \mathcal{P}_k\} \cap I| \leqslant |\mathcal{I}| + |\mathcal{P}_k \setminus \mathcal{I}| = |\mathcal{P}_k| = 2^{2^{k-1}-1},$$
$$\tilde{w}_k(\tilde{G}_k) = w_k(G_k) + |\mathcal{P}_k| = \tfrac{k+3}{2} \cdot 2^{2^{k-1}-1}.$$

The graph $\tilde{G}_k$ is the smallest one for which I can prove that it has a weight assignment such that the ratio between the maximum weight of an independent set and the total weight is at most $\frac{2}{k+3}$ It is not difficult to prove (e.g. using weak LP duality) that the assignment of weights $\tilde{w}_k$ to the vertices of $\tilde{G}_k$ is optimal (gives the least ratio) for this particular graph.

Both constructions give rise to triangle-free intersection graphs of n segments in the plane with maximum independent set size $\Theta(n/\log\log n)$. On the other hand, it follows from the result of McGuinness [4] that every triangle-free intersection graph of n segments has chromatic number $O(\log n)$ and maximum independent set size $\Omega(n/\log n)$.

4 OTHER GEOMETRIC SHAPES

It is known that the graphs G_k and $\tilde{G}_k$ have intersection models by many other geometric shapes, for example, L-shapes, axis-parallel ellipses, circles, axis-parallel square boundaries [6] or axis-parallel boxes in $\mathbb{R}^3$ [2]. The result of this paper can be extended to those models for which every geometric object X representing a vertex of the intersection graph can be replaced by many pairwise disjoint objects approximating X. This is possible, for example, for intersection graphs of L-shapes, circles or axis-parallel square boundaries, but not for intersection graphs of axis-parallel ellipses or axis-parallel boxes in $\mathbb{R}^3$. The problem whether triangle-free intersection graphs of the latter kind of shapes have linear-size independent sets remains open.

ACKNOWLEDGMENTS

The author thanks Michael Hoffmann for helpful discussions. The work was supported by Polish National Science Center Grant 2011/03/B/ST6/01367 and Swiss National Science Foundation Grant 200020-144531.

REFERENCES

1. Ackerman, E.: On the maximum number of edges in topological graphs with no four pairwise crossing edges. Discrete Comput. Geom. **41**(3), 365–375 (2009)

2. Burling, J. P.: On coloring problems of families of prototypes. Ph.D. Thesis, University of Colorado, Boulder (1965)

3. Fox, J., Pach, J.: Coloring *K k* Kk -free intersection graphs of geometric objects in the plane. Eur. J. Comb. **33**(5), 853–866 (2012)

4. McGuinness, S.: Colouring arcwise connected sets in the plane I. Graphs Comb. **16**(4), 429–439 (2000)

5. Pach, J., Shahrokhi, F., Szegedy, M.: Applications of the crossing number. Algorithmica 16(1), 111–117 (1996)

6. Pawlik, A., Kozik, J., Krawczyk, T., Lasoń, M., Micek, P., Trotter, W.T., Walczak, B.: Triangle-free geometric intersection graphs with large chromatic number. Discrete Comput. Geom. **50**(3), 714–726 (2013)

7. Pawlik, A., Kozik, J., Krawczyk, T., Lasoń, M., Micek, P., Trotter, W.T., Walczak, B.: Triangle-free intersection graphs of line segments with large chromatic number. J. Comb. Theory, Ser. B **105**, 6–10 (2014)

CHAPTER 3

Weighted Straight Skeletons in the Plane

Therese Biedl[a], Martin Held[b], Stefan Huber[c], Dominik Kaaser[b], Peter Palfrader[b]

[a] *David R. Cheriton School of Computer Science, University of Waterloo, Waterloo, Ontario N2L 1A2, Canada*

[b] *Universität Salzburg, FB Computerwissenschaften, 5020 Salzburg, Austria*

[c] *Institute of Science and Technology Austria, 3400 Klosterneuburg, Austria*

ABSTRACT

We investigate weighted straight skeletons from a geometric, graph-theoretical, and combinatorial point of view. We start with a thorough definition and shed light on some ambiguity issues in the procedural definition. We investigate the geometry, combinatorics, and topology of faces and the roof model, and we discuss in which cases a weighted straight skeleton is connected. Finally, we show that the weighted straight skeleton of even a simple polygon may be non-planar and may contain cycles, and we discuss under which restrictions on the weights and/or the input polygon the weighted straight skeleton still behaves similar to its unweighted counterpart. In particular, we obtain a non-procedural description and a linear-time construction algorithm for the straight skeleton of strictly convex polygons with arbitrary weights.

Keywords: Straight skeleton; Generalization; Characterization; Ambiguity; Positive and negative weights

1. INTRODUCTION

The straight-skeleton S(P)S(P) of a simple polygon P is a skeleton structure that was introduced to computational geometry by Aichholzer et al. [1] about 20 years ago. Its definition is based on a wavefront propagation process where the polygon's edges move inwards at unit speed. The straight skeleton, roughly speaking, is the skeleton structure that results from the interference patterns of the wavefront edges. Aichholzer and Aurenhammer [2] later generalized the definition to planar straight-line graphs. Since their introduction, a lot of applications appeared in different research areas, and multiple algorithms to compute the straight skeleton have been introduced [3].

Eppstein and Erickson [4] were the first to mention the weighted straight skeleton where the wavefront edges may move with arbitrary but fixed speeds. They claim that their algorithm to compute the unweighted straight skeleton in $O(n^{8/5+\epsilon})$ time and space also works, without major changes, for weighted straight skeletons. Weighted straight skeletons have many applications: Barequet et al. [5] use weighted straight skeletons in order to define the initial wavefront topology for straight skeletons of polyhedra. Haunert and Sester [6] use the weighted straight skeleton for topology-preserving area collapsing in geographic maps. Laycock anDay [7] and Kelly and Wonka [8] use weighted straight skeletons to model realistic roofs of houses. Aurenhammer [9] investigated fixed-share decompositions of convex polygons using weighted straight skeletons with specific positive weights. Sugihara [10] employs weighted straight skeletons for the interactive design of pop-up cards.

Although algorithms, applications, and even simple implementations [11] of weighted straight skeletons are known, only limited research has been conducted on the weighted straight skeleton itself. The only known results are that the simple definition based on wavefront propagation may lead to ambiguities [3] and [8] and that the lower envelope characterization by Eppstein and Erickson [4] does not apply. In this paper, we carefully define weighted straight skeletons, shed light on the ambiguity in the procedural definition, investigate geometric, graph-theoretical and combinatorial properties of weighted straight skeletons, and compare those with properties of unweighted straight skeletons. In particular, we show that weighted straight skeletons of simple polygons may have cycles and crossings. Furthermore, we investigate necessary conditions for the weights or the polygon such that the weighted straight skeleton of a simple polygon is a planar tree.

2. PRELIMINARIES

The definition of the straight skeleton S(P) of a simple polygon P is based on a so-called wavefront propagation of P where all edges of P move inwards in parallel and at unit speed. (This definition is readily extended to polygons with holes.) The wavefront, denoted by WP(t), has the shape of a mitered offset curve of P for small t. As t increases, WPWP changes its topology. Such changes are called events and we can distinguish between two main types: An edge event happens when an edge e collapses to zero length and vanishes. A split event happens when a reflex wavefront vertex v reaches another part of the wavefront and splits it into parts. Typically, v meets a wavefront edge whose split causes the entire wavefront to split into two parts. However, if v meets one or more other wavefront vertices, then more complicated splits of the wavefront into multiple parts are possible. Either event causes local changes in the topology of the wavefront so that the resulting wavefront again consists of a collection of simple polygons.

The straight skeleton S(P)S(P) is defined as the set of loci traced out by the vertices of WP(t) for all t≥0, see **Figure**. 1. Additionally, some loci are added to the straight skeleton in case of parallel edges as follows: (a) If two parallel edges e and e′e′ that move in opposite directions become overlapping during an event, then the region common to e and e′e′ is added to the straight skeleton, while the region(s) that belongs to exactly one of them remains in the wavefront. (b) If two parallel edges e and e′e′ that move in the same direction become adjacent due to an edge event, then their common endpoint is considered a vertex of the wavefront. We call this a ghost vertex. This vertex moves perpendicular to e and e′e′.1

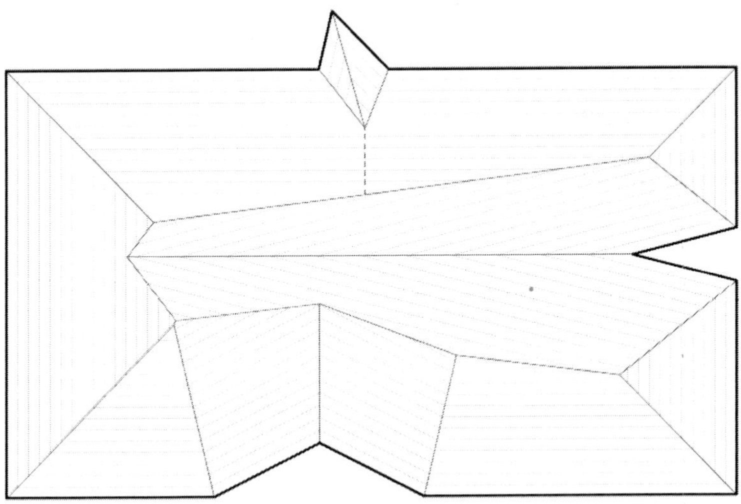

Figure 1. The straight skeleton S(P)S(P) of the input polygon P (bold) is defined by wavefronts emanated from P. The dashed arc is traced by a ghost vertex.

Each wavefront vertex traces out an arc of S(P)S(P). Since wavefront vertices move along bisectors of edges of P , the arcs of S(P)S(P) are straight-line segments. Every event of WPWP corresponds to a locus where arcs of S(P)S(P) meet and give rise to a node of S(P). See **Figure**. 1 for an example. The straight skeleton S(P)S(P) is interpreted as a graph, and one can show that it is a tree [1]. Also, no two arcs of the straight skeleton cross since the wavefront moves inwards towards the unswept region. Hence, S(P)∪PS(P)∪P is a planar straight-line graph. The inner faces of S(P)∪PS(P)∪P are called straight-skeleton faces.

Let e be an edge of a polygon P, and let the wavefront fragments of e at time t be the union of segments of WP(t) that originated from e . We denote this set by e(t), and it may comprise none, one, or many segments, depending on whether e participated in edge events and/or split events. We consider segments in e(t) to be open line segments. Every straight-skeleton face f(e)f(e) is traced out by (the fragments of) one wavefront edge e , thus f(e):=∪t>0e(t)f(e):=∪t>0e(t). It can be shown that f(e)f(e) is monotone with respect to the supporting line of e [1] and [2], and its lower chain is convex [3]. Furthermore, the boundary of each face f(e) corresponds to a cycle in P∪S(P)P∪S(P).

2.1. Roof models

Aichholzer et al. [1] introduced the roof model, which is a handy way of interpreting the straight skeleton. One considers the wavefront propagation embedded in three-space where the z -axis constitutes time. Then WPWP traces out a surface, namely the roof model $\mathcal{T}(P) := \bigcup_{t \geq 0} \mathcal{W}_P(t) \times \{t\}$. This concept gives us the means to investigate WP over its entire lifespan. Note that we can obtain S(P)S(P) from T(P)T(P) by projecting the edges of T(P)T(P) onto the plane $\mathbb{R}^2 \times \{0\}$. Conversely, we can obtain T(P) from S(P) by lifting all nodes of S(P) by their orthogonal distance to the respective input edges of P. These distances correspond to the times when the nodes were swept by the wavefront. The roof model is sometimes also called terrain model since T(P)T(P) is a terrain, i.e., any line parallel to the z-axis intersects it in at most one point. As this property may be violated for the weighted version of straight skeletons, we prefer the term "roof model".

2.2. Weighted straight skeletons

The weighted straight skeleton differs from the (unweighted) straight skeleton only in the speed σ(e)∈R\{0}σ(e)∈R\{0} with which an edge e moves in the wavefront. We call σ the weight function and σ(e)σ(e) the weight of e . The wavefront moves such that the fragments e(t)e(t) of e at time t are on the line $\bar{e} + t \cdot \sigma(e) \cdot n(e)$is the supporting line of e , and n(e)n(e) is the inward unit normal of e . For σ(e)<0σ, an edge e moves outward with speed |σ(e)|.2

The definitions of edge event, WP(t)WP(t), S(P)S(P), e(t)e(t), f(e)f(e), and T(P)T(P) for the unweighted straight skeleton carry over verbatim to the weighted straight skeleton. We use an additional parameter σ in WP(t,σ)WP(t,σ), S(P,σ)S(P,σ), and T(P,σ)T(P,σ) in order to refer to their weighted counterparts. The major difference, however, lies in the richness of the notation of a split event: A split event happens when the wavefront becomes non-simple because a wavefront vertex v meets another part of the wavefront. The vertex v no longer needs to be reflex. Also, a split event does not necessarily split the wavefront into multiple parts. Even worse, different possibilities may exist on how to proceed with the wavefront propagation such that the wavefront still remains a planar collection of simple polygons between events. We discuss this in more detail in Section 4.

Negative weights may cause wavefront components to propagate towards infinity. The corresponding roof model is unbounded in this case. From a combinatorial point of view, we add a node at infinity for every wavefront component C that still exists at a sufficiently large time such that all events already have been processed. Each such node is incident to the infinite arcs that are traced out by vertices of C . Doing so gives us the important property that the boundary of each face f(e)f(e) is a cycle in P∪S(P,σ)P∪S(P,σ) even for unbounded faces.

While most definitions carry over, it is not at all clear which of the properties of the straight skeleton, such as planarity, connectedness, acyclicity, or monotonicity of faces, remain valid in the weighted version. This is the main topic of this paper and will be discussed in Section 5.

3. VELOCITY OF WAVEFRONT VERTICES

For later argumentations, we first develop a formula that describes how the movement of a wavefront vertex v is determined by the propagation of the two incident wavefront edges e1e1 and e2e2 which we assume to be not collinear. (See Section 4 for a discussion of degeneracies.) We use σ(v)∈R2σ(v)∈R2 for the velocity vector of v.

Notice that σ(v)σ(v) lies in one of the four sectors spanned by the supporting lines of e1e1 and e2e2, and we know which one of these to use from the signs of both σ(e1)σ(e1) and σ(e2)σ(e2). Therefore, we can determine which edge among e1,e2e1,e2 is the first edge in counter-clockwise direction from σ(v)σ(v) and call this the left edge of v. Similarly, we determine the first edge in clockwise direction and call this right edge of v . After possible renaming, assume that e1e1 is the left edge and e2e2 is the right edge as shown in **Figure**. 2. We denote by α1α1 the counter-clockwise angle between σ(v)σ(v) and e1e1 and by α2α2 the clockwise angle between σ(v)σ(v) and e2e2.

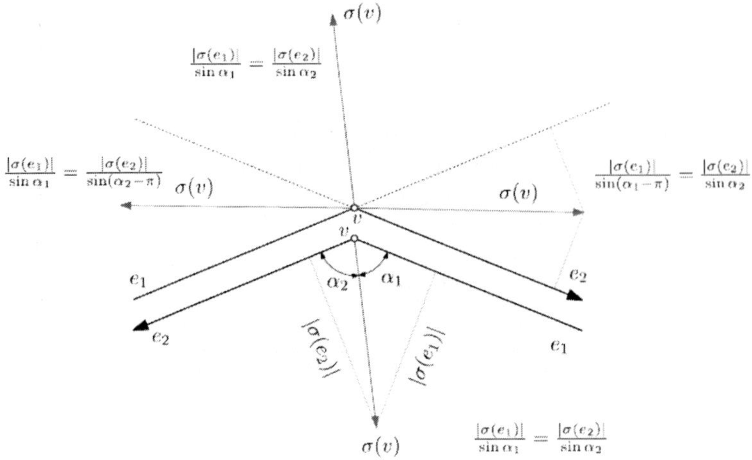

Figure 2. The velocity vector σ(v) of a wavefront vertex v with left edge e1e1 and right edge e2e2 lies in one of four sectors spanned by the supporting lines of e1e1 and e2e2.

We have multiple cases to consider, based on the signs of σ(e1)σ(e1) and σ(e2)σ(e2) and on whether v is convex or reflex. With our choice of naming for α1α1 and α2α2 dependent on the wedge containing σ(v)σ(v), these cases can be combined, and we can express ||σ(v)||||σ(v)|| by |σ(ei)||σ(ei)| and αiαi as shown in **Figure**. 2 as follows:

$$\left\|\sigma(v)\right\| = \begin{cases} \dfrac{|\sigma(e_i)|}{\sin\alpha_i} & \text{for } \alpha_i < \pi \\ \dfrac{|\sigma(e_i)|}{\sin(\alpha_i - \pi)} & \text{for } \alpha_i > \pi \end{cases}, \tag{1}$$

for all i∈{1,2}i∈{1,2}. Note that αi=π is impossible as σ(ei)≠0. In case that αi>π, we have $\operatorname{sign}(\sigma(e_1)) \neq \operatorname{sign}(\sigma(e_2))$. Using sin(αi−π)=, it follows that in all cases

$$\frac{\sigma(e_1)}{\sin\alpha_1} = \frac{\sigma(e_2)}{\sin\alpha_2}. \tag{2}$$

Denoting by $\gamma = \alpha_1 + \alpha_2$ the angle spanned by e_1 and e_2 on the side where v moves to, we see that

$$\frac{\sin\alpha_1}{\sigma(e_1)} = \frac{\sin(\gamma - \alpha_1)}{\sigma(e_2)} = \frac{\sin\gamma\cos\alpha_1 - \cos\gamma\sin\alpha_1}{\sigma(e_2)},$$

and hence we obtain

$$\cot\alpha_1 = \frac{\cos\gamma + \frac{\sigma(e_2)}{\sigma(e_1)}}{\sin\gamma}. \tag{4}$$

From the case distinctions on sign(σ(e_1))sign(σ(e_1)) and sign(σ(e_2))sign(σ(e_2)) we can uniquely determine α1 and, using Eq. (1), obtain ||σ(v)|| ||σ(v)||.

4. AMBIGUITY OF DEFINITION

4.1. Ambiguity at edge events

Presume that we have an edge event where edge e0 disappears, leaving its adjacent edges e_1 and e_2 to become adjacent with common endpoint v as shown in **Figure**. 3.

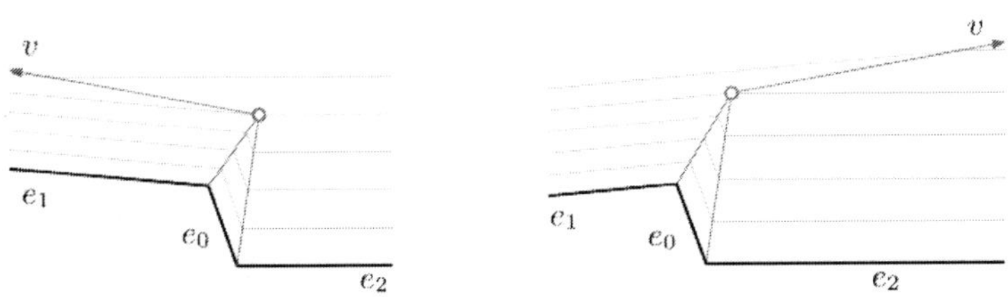

Figure 3. The definition of straight skeletons is ambiguous when two parallel wavefront edges with different weights become adjacent.

Recall that Eq. (3) describes the angle at which vertex v moves after the event. However, the right-hand side of this equation is indeterminate for sin(γ)=0sin(γ)=0, i.e., when e1e1 and e_2e_2 are parallel. Unfortunately, as γ monotonically approaches π, we may obtain two different limit cases:

$$\lim_{\gamma \nearrow \pi} \alpha_1 = 0 \qquad \lim_{\gamma \searrow \pi} \alpha_1 = \pi .$$

(4)

Therefore, and as already alluded to by Kelly and Wonka [8] and Huber [3], the definition of the weighted straight skeleton is ambiguous whenever parallel wavefront edges with different weights become adjacent. In the case where σ(e_1)=σ(e_2) we can apply de l'Hôpital's rule to obtain lim $_{\gamma \to \pi} \cot \alpha_1 = 0$, and hence $\alpha_1 = \pi/2$. Similarly, $\alpha_1 = \pi \pm \pi/2$ if $\sigma(e_1) = -\sigma(e_2)$ But in all other cases, there is no unique definition of a straight skeleton.

Whenever two edges become adjacent in the ambiguous case, there are different options on how to proceed. At first sight, it appears possible to treat this case equal to the well-defined case and keep propagating both edges. Unfortunately, regardless of which direction we choose for the bisector of these two edges, we will run into a problem: the difference in speed of edge propagation results in an ever growing gap between the two parts of the wavefront made up by e1e1 and

e_2. This gap could be filled by creating a new wavefront edge that again connects e_1 and e_2. However, this approach seems inadvisable: This new wavefront edge would not belong to any specific input edge and it would not necessarily be parallel to any input edge either. Also, depending on the exact definition of this new edge, most likely it would have a propagation speed of zero, resulting in a degenerate face.

We suggest that the only reasonable way of handling the ambiguity is to terminate propagation of one of the edges in favor of the other. The limit cases suggested by Eq. (4) support this approach as they set α to either 0 or π. It is unclear which of the two limits one should choose as both choices appear equally justified. To get a deterministic algorithm, one canonical strategy might be to always pick the faster, or always the slower, of the two edges to continue its propagation. The first choice emphasizes the idea of the faster wavefront edge overtaking the slower one and leads to the nice property that the arc traced out by v is convex in the roof model. The second choice increases the slope of the roof and, thus, the volume under the roof.

Let the left edge e_1 be the one we decided to stop. We add e_1 to S(P). Then, edge e_2 is enlarged to cover the wavefront previously belonging to e_1 and continues its propagation at its defined speed.

This ambiguity can already arise at time t=0, i.e., for the input polygon: Consider a polygon with two consecutive edges that are collinear. Unless the weights of these two edges are identical, we end up with one face of S(P) that has zero area.

Note that the situation can become even more complicated if multiple parallel wavefront edges with different weights become consecutive, as depicted in **Figure**. 4. Like in the simple case, there are many equally justified ways to continue the wavefront propagation. For the sake of simplicity, we will describe just one of them.

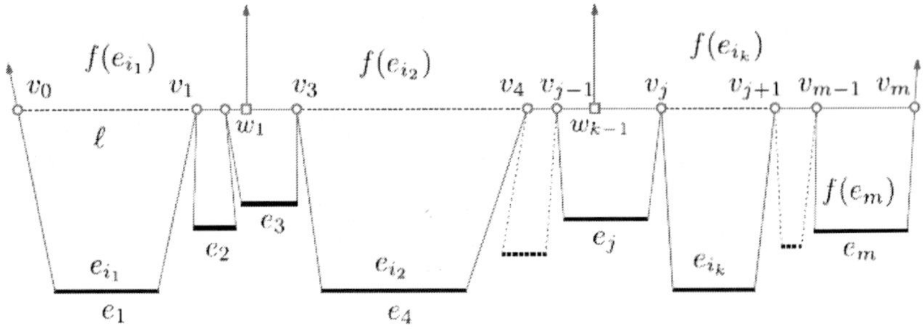

Figure 4. Continuing the wavefront propagation if multiple parallel wavefront edges become adjacent when simultaneously reaching the line ℓ.

Assume, as in **Figure**. 4, that at some point in time a line ℓ contains part of the wavefront. Let a maximal contiguous part of the wavefront on ℓ consist of vertices and edges $v_0, e_1, v_1, \ldots, e_m, v_m$ (in order). Also assume for now that $e_1, \ldots, e_m$ em all approach ℓ from the same half-plane. Let $1 \leq i_1 < \cdots < i_j < \cdots < i_k \leq m$ be indices such that edges $e_{i_1}, \ldots, e_{i_k}$ are those edges with maximal weight, i.e., $\sigma(e_{i_1}) = \cdots = \sigma(e_{i_k}) = \max_{1 \leq i \leq m} \sigma(e_i)$. We now modify the wavefront such that those edges take over the wavefront while all other edges disappear. For each $j = 1, \ldots, k-1$, we create a ghost vertex wjwj halfway between vijvij and $v_{i_{j+1}-1}$, i.e., we add ghost vertices between all edges of maximal speed. Then, we extend each of the edges ei1,...,eikei1,...,eik on both sides to their respective ghost vertices (or to v0v0 and vmvm at the ends) anDrop all other edges. This replaces the wavefront between v_0 and $v_m v_m$ with edges $(v_0, w_1), (w_1, w_2), \ldots, (w_{k-2}, w_{k-1}), (w_{k-1}, v_m)$. Thus, these edges take over this part of the wavefront from all other edges and continue to propagate while the remaining edges disappear.

If not all edges on ℓ approach from the same half-plane, then any non-empty line segment that is common to two edges in opposite direction becomes an arc of the weighted straight skeleton. We otherwise proceed as above: The edge(s) with the maximum speed take over all other adjacent edges of the wavefront that reside on ℓ.

4.2. Ambiguities at coinciding split events

A second ambiguity may arise when multiple split events happen at the same time and at the same locus p of the wavefront. How should the topology of the wavefront change? It should be done in such a way that afterwards the wavefront again consists of a planar collection of simple polygons, i.e., so that the wavefront near p has no crossing after the event.

Denote by e1,...,e2ke the edges that end at p when the event happens, see **Figure**. 5. We say eiei and ejej are paired if the wavefront near p consisted of edge eiei, then (for the pairing before the event) possibly a few edges that vanish during the event at p , and then edge ejej. For a specific pairing before the event, we are interested in a valid pairing after the event such that the wavefront remains crossing-free.

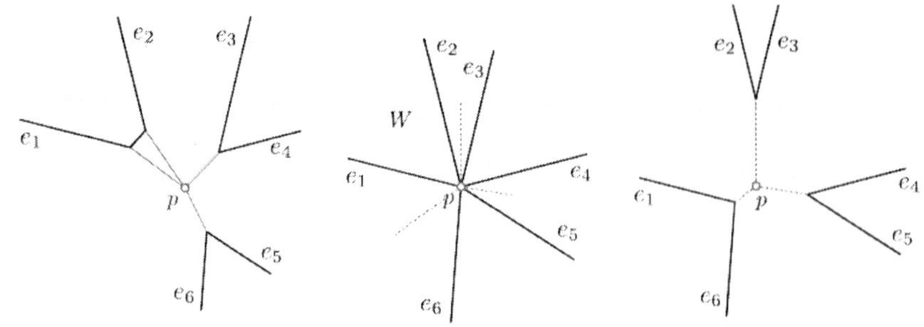

Figure 5. Three wavefront vertices meet simultaneously at p . Left: Before the event the pairing of wavefront edges is $(e_1, e_2), (e_3, e_4), (e_5, e_6)$. Middle: At the event, the edges tessellate the neighborhood of p into interior-disjoint wedges. Every second wedge was already swept by the wavefront. Right: After the event the pairing changes to $(e_2, e_3), (e_4, e_5), (e_6, e_1)$.

Roughly speaking, if e1,…,e2ke1,…,e2k are sorted clockwise around p and the pairing before the event was $(e_1, e_2), (e_3, e_4), \ldots, (e_{2k-1}, e_{2k})$, then we define the pairing $(e_2, e_3), (e_4, e_5), \ldots, (e_{2k}, e_1)$ as the standard resolution, see **Figure**. 5.

Lemma 1. For positive weights the standard resolution always gives a valid pairing and, moreover, it is the only valid solution.

Proof. We first remark that for positive weights a locus is never swept twice by the wavefront, as we show in Section 5.1. Furthermore, the wavefront is planar until it reaches p. The essential observation is the following: When the wavefront reaches p, the neighborhood of p is tessellated into interior-disjoint wedges bounded by e1,…,e2ke1,…,e2k, and every other wedge is completely swept by the wavefront. Assume that e1,…,e2ke1,…,e2k are sorted clockwise around p and the wedge W between e1,e2e1,e2 is already swept. That means that before the event the edges e1e1 and e2e2 resided in W . As the wedges are interior-disjoint, e1e1 and e2e2 were paired before the event. By analogous arguments for the other edges, (e1,e2),(e3,e4),…,(e2k−1,e2k)(e1,e2),(e3,e4),…,(e2k−1,e2k) was the pairing before the event.

The standard resolution alters the pairing to (e2,e3),(e4,e5),…,(e2k,e1)(e2,e3),(e4,e5),…,(e2k,e1). Observe that all edge pairs e2ie2i and $e_{2(i \bmod k)+1}$<img height="12" border="0" style="vertical-align:bottom" width="74" alt="View the MathML source" title="View the MathML source" src="http://origin-ars.els-cdn.com/content/image/1-s2.0-S0925772114000807-si87.gif">e2(imodk)+1, with 1≤i≤k1≤i≤k, reside in different unswept wedges, which proves the validity of the standard resolution. Assume now that there is an alternative pairing. We may assume that e2e2 is paired with an edge other than e3e3. But then e2e2 would necessarily cross an already swept area which is impossible for positive weights. □

For negative weights, it is not at all obvious why a pairing without crossings should even exist. **Figure**. 6 shows an example where the standard pairing can not be applied, as it results in crossings.

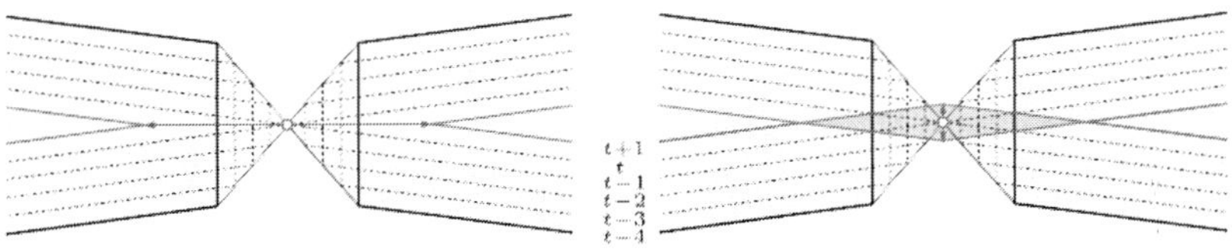

Figure 6. Two edge events (for the vertical edges) and multiple split events coincide at time t. We must retain the pairing as it was after vanishing edges have been excised; otherwise crossings result.

We can transform the problem of finding a suitable pairing to a stable roommates problem, as defined by the order of intersections along the edges. This stable roommate problem can be shown to possess a stable matching. However, the details of this technique and the proofs are non-trivial and will be covered in a separate paper. (A preliminary version appeared as Biedl et al. [12].)

Another difficulty is that there may be multiple valid pairings that have no crossings.

Lemma 2. {+1,−1}{+1,−1}, for which multiple post-event wavefronts without crossings exist.

Proof. The example is shown in **Figure**. 7.

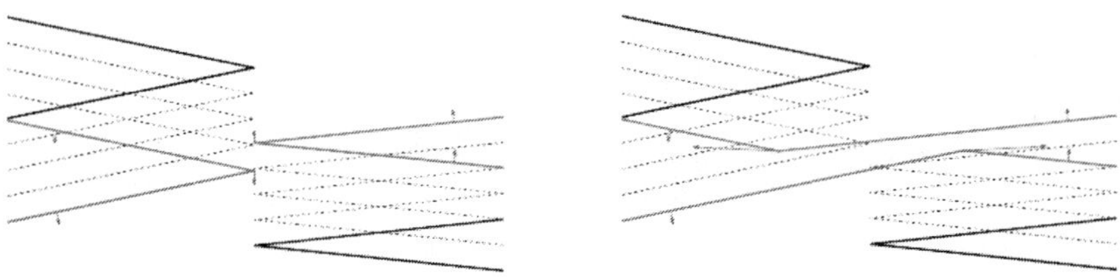

Figure 7. When multiple reflex vertices meet at an event, there may be multiple possible new wavefronts. Different wavefronts after the event are shown in the **Figure**eures on the left and on the right. The old pairing (left) and the standard pairing (right) both yield simple wavefronts.

5. PROPERTIES OF STRAIGHT SKELETONS

For the sake of simplicity and clarity, we assume in the following that the ambiguities discussed in Section 4 do not occur. That means we assume that (i) there are no two parallel edges with different weights that become consecutive due to an edge event or that are consecutive initially, and (ii) that two or more events do not happen simultaneously at the same place unless they are all edge events. The latter implies that split events are exclusively caused by vertices reaching the interior of an edge of the wavefront. But, as we show now, even under these assumptions many seemingly natural properties do not hold for weighted straight skeletons. Table 1 summarizes all results.

Table 1. Results for a polygon with h holes, a simple polygon, and a convex polygon, with n denoting the number of vertices of the polygon P and v denoting the number of nodes of *S*(P)

POLYGON WITH HOLES	$\sigma \equiv 1$	σ positive	σ arbitrary
$S(P)$ is connected	✓ Lemma 15	✓ Lemma 15	× Lemma 12
$S(P)$ has no crossing	✓ Lemma 6	✓ Lemma 6	× Lemma 3
$f(e)$ is monotone w.r.t. e	✓ as in [1]	× Lemma 11	× Lemma 11
bd $f(e)$ is a simple polygon	✓ as in [1]	× Lemma 8	× Lemma 8
$T(P,\sigma)$ is z-monotone	✓ Lemma 4	✓ Lemma 4	× Lemma 5
$S(P)$ has $n+v-1+h$ arcs	✓ Corollary 17	✓ Corollary 17	× Lemma 3
$S(P)$ is a tree	× Corollary 17	× Corollary 17	× Lemma 12
SIMPLE POLYGON	$\sigma \equiv 1$	σ positive	σ arbitrary
$S(P)$ is connected	✓ [1]	✓ Lemma 15	✓ Corollary 14
$S(P)$ has no crossing	✓ [1]	✓ Lemma 6	× Lemma 3
$f(e)$ is monotone w.r.t. e	✓ [1]	× Lemma 11	× Lemma 11
bd $f(e)$ is a simple polygon	✓ [1]	✓ Lemma 9	× Lemma 10
$T(P,\sigma)$ is z-monotone	✓ [1]	✓ Lemma 4	× Lemma 5
$S(P)$ has $n+v-1$ arcs	✓ [1]	✓ Corollary 17	× Lemma 3
$S(P)$ is a tree	✓ [1]	✓ Corollary 19	× Lemma 3
CONVEX POLYGON	$\sigma \equiv 1$	σ positive	σ arbitrary
$S(P)$ is connected	✓ [1]	✓ Lemma 15	✓ Corollary 14
$S(P)$ has no crossing	✓ [1]	✓ Lemma 6	× Lemma 20
$f(e)$ is monotone w.r.t. e	✓ [1]	✓ Corollary 25	✓ Corollary 25
bd $f(e)$ is a simple polygon	✓ [1]	✓ Lemma 9	✓ Corollary 25
$T(P,\sigma)$ is z-monotone	✓ [1]	✓ Lemma 4	✓ Theorem 24
$S(P)$ has $n+v-1$ arcs	✓ [1]	✓ Corollary 17	✓ Corollary 29
$S(P)$ is a tree	✓ [1]	✓ Corollary 19	✓ Lemma 28

Lemma 3. There exists a simple polygon P with weights chosen from the $\{+1, -1\}$ such that $S(P,\sigma)$ has crossings and cycles.

Proof. The polygon is shown in **Figure**. 8.

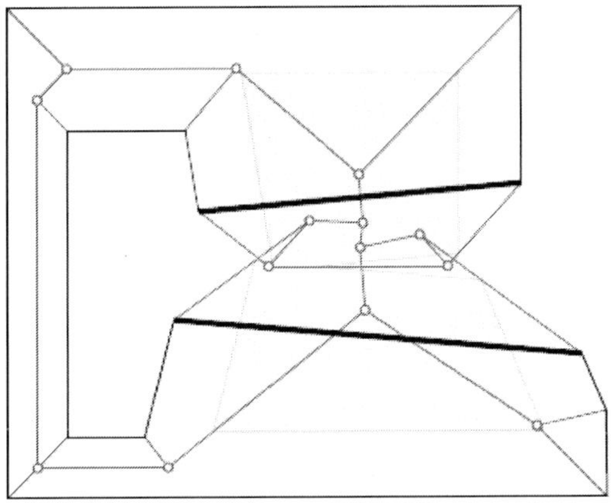

Figure 8. S(P,σ)S(P,σ) of a simple polygon P may have crossings and cycles. A wavefront at some time t is shown in gray. All edges have weight +1, except the two bold edges, which have a negative weight of −1.

5.1. Terrains and crossings

We now want to show that if the weights are positive, then the straight skeleton has no crossing. For this (and other claims later) it will help to study when the roof model T(P,σ)T(P,σ) is a terrain.

Lemma 4. Let P be a polygon (possibly with holes). If $\sigma(e) > 0$ for all wavefront edges e, then the roof model $T(P,\sigma)$ is a terrain and its z-projection equals P.

Proof. Since all weights are positive, the wavefront edges emanated by P move towards the interior of P . Therefore, after each event of WP,σWP,σ, the trajectories of the newly born wavefront vertices point to the area within P that has not yet been swept by the wavefront. Thus, no wavefront vertex can ever reach a locus that has already been swept—such a vertex would have met another part of the wavefront before and the vertex would have been annihilated. Accordingly, the wavefront WP,σWP,σ stays within P and no locus of P is swept more than once by the wavefront. On the other hand, each locus of P is swept at least once, since otherwise the boundary of the unswept region would be the wavefront, and therefore not yet empty.

Note that Lemma 4 fails to hold as soon as there is a negative weight as the following lemma shows.

Lemma 5. There exists a simple polygon P with weights chosen from the set $\{+1, -1\}$ such that $\mathcal{T}(P, \sigma)$ is not a terrain.

Proof. The example shown in **Figure**. 8 contains loci that are swept more than once. Hence, T(P,σ)T(P,σ) is not a terrain. In fact, it can be modified easily such that some loci are swept an arbitrary number of times.

Lemma 6. Let P be a polygon (possibly with holes). $\sigma(e) > 0$ for all wavefront edges e, then $\mathcal{S}(P, \sigma)$ has no crossings.

Proof. This holds by Lemma 4 since the locus p of any crossing must have been covered at least twice by the wavefront. But then the line parallel to the z-axis through p would intersect T(P,σ)T(P,σ) twice.

We provide one more result, showing that, while weighted straight skeletons lack many of the properties of unweighted straight skeletons, their terrains are, nevertheless, not entirely arbitrary. The next lemma generalizes a theorem by Aichholzer et al. [1] that states that rain does not cumulate on roofs stemming from straight skeletons; instead raindrops can run off. In this sense, we define a pit of $\mathcal{T}(P, \sigma)$ as a point p on $\mathcal{T}(P, \sigma)$ for which no z-monotone path from p to P on $\mathcal{T}(P, \sigma)$) exists.

Lemma 7. Let P be a polygon (possibly with holes) and let σ be an arbitrary weight function. Then $\mathcal{T}(P, \sigma)$ has no pit.

Proof. Consider a point p on the roof and say it was reached at time t>0t>0. Let p lie in the closure of the face f(e)f(e) swept by wavefront fragments emanated from the input edge e. This face was traced out during the wavefront propagation process, i.e., there is a (not necessarily strictly) t-monotone path from e to p. Thus, in the roof, there is a z-monotone path ascending from e at z=0z=0 to p. Therefore, no point p of the roof can be a pit.

5.2. Straight-skeleton faces

We later want to argue that under some assumptions the straight skeleton is connected. To do so, we first study some properties of the faces of edges. Recall that $f(e) = \bigcup_{t>0} e(t)$, where $e(t)$ are the open line segments that result from edge e at time t . Clearly, $f(e)$ is connected (no fragment of e suddenly appears during a wavefront process) and its boundary $\mathrm{bd}\, f(e)$ consists of arcs of the straight skeleton and the input edge e.

Lemma 8. There exists a polygon P with holes and with positive weights chosen from the set $\{1, 3\}$ such that $\mathrm{bd}\, f(e)$ is not a simple polygon.

Proof. Polygon P is shown in **Figure**. 9 (include the dotted features), with edge e being the bottom-most horizontal edge. During the wavefront process, edge e gets split when it meets the hole. But since e moves faster than the edges of the hole, two fragments of e later re-combine. According to our definition of straight skeleton, a ghost vertex is created that traces the arc a in **Figure**. 9. The boundary of $f(e)$, viewed as a polygon, is not simple and can be interpreted to contain the arc a twice. (Even without this arc, the boundary $\mathrm{bd}\, f(e)$ would not be one simple polygon.)

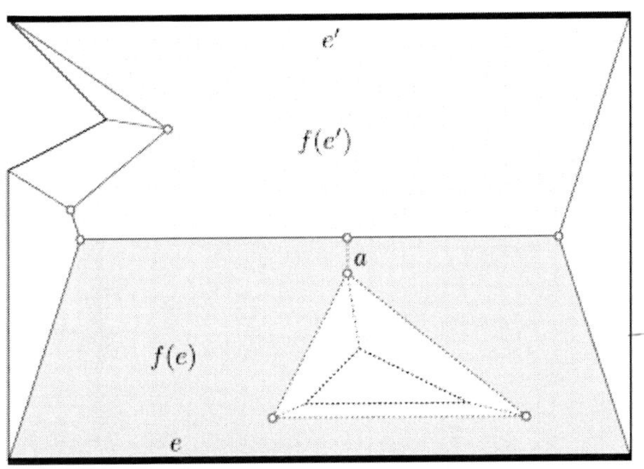

Figure 9. A polygon with hole may have a non-simple face. A simple polygon may have a non-monotone face. The dotted features are caused by the hole. The bold edges have weight 3, the others have weight 1.

Lemma 8 used a polygon with holes. Whether for simple polygons a similar situation can arise depends on whether the weights are all positive or not.

Lemma 9. Let P be a simple polygon and let σ be an assignment of positive weights to the edges of P. Then $\mathrm{bd}\, f(e)$ is a simple polygon for all edges e.

Proof. Note that $f(e)$ is an open, simply-connected set: Whenever a fragment splits, the pieces never re-merge due to the ghost vertices. Hence $\mathrm{bd}\, f(e)$ is a weakly-simple3 polygon, and the only way that it could be not simple is by having a point p that is incident to three or more edges of the boundary of $f(e)$; see **Figure**. 9 around a . This can only happen if two fragments s_1, s_2 in $f(e)$ become adjacent in the wavefront. By tracing from s_1 and s_2 back to e while residing inside $f(e)$, we can find a closed Jordan curve C inside $\mathrm{cl}\, f(e)$ that encloses a point outside $\mathrm{cl}\, f(e)$

For positive weights the roof model $T(P)$ projects to $C \subset \mathrm{cl}\, f(e) \subset P$. Since P is simple, any point inside C also belongs to P. Since some point inside C does not belong to $\mathrm{cl}\, f(e)$ some other face $f(e)$ is inside C. But by planarity (Lemma 6) the edge e′ of this face is inside C as well. Since C contains no edges of P , edge e′ cannot be connected to the edges at the exterior face of P. So P has a hole, a contradiction.

Lemma 10. There exists a simple polygon P with weights chosen from the set $\{-1, +1, +3\}$ }such that $\mathrm{bd}\, f(e)$ is not a simple polygon for an edge e of P.

Proof. The polygon is shown in **Figure**. 10.

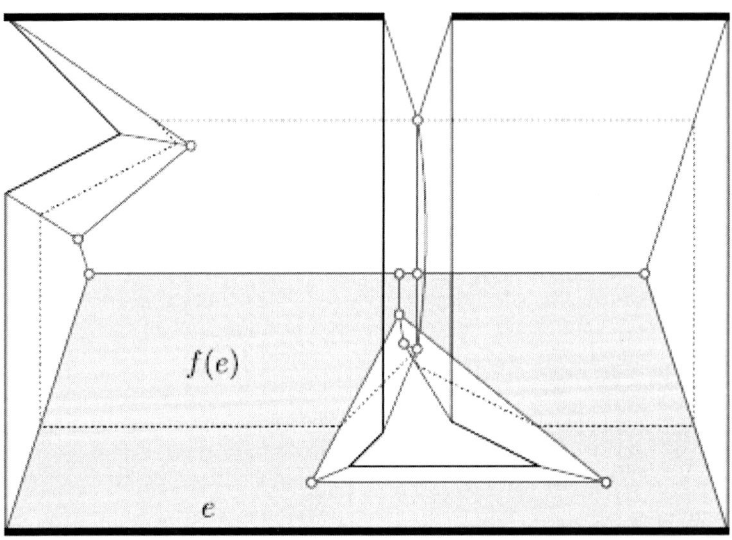

Figure 10. A simple polygon where face $f(e)$ (shaded) has a non-simple boundary. One wavefront is depicted by dotted lines. The bold edges have weight 3 and the two vertical edges that form the corridor have weight −1. All other edges have weight 1. The arc between the two corridor edges geometrically coincides with other arcs.

If all weights are the same, then $\mathrm{bd}\, f(e)$ is simple even for a polygon with holes, because $f(e)$ is monotone with respect to the supporting line of e [1]. We note here that $f(e)$ need not be monotone if we allow different weights. This is already obvious from **Figure**. 9 and **Figure**. 10 but may happen even for simple polygons and positive weights.

Lemma 11. There exists a simple polygon P with weights chosen from the set {1,3} such that for one edge e the face $f(e)$ is not monotone with respect to the supporting line of e.

Proof. The face f(e′)f(e′) in **Figure**. 9, with the dotted features omitted, is not monotone.

5.3. Connectivity

The unweighted straight skeleton $\mathcal{S}(P)$ of a polygon P with holes is always connected. In fact, $\mathcal{S}(P)$ is even of the same homotopy type as P. The weighted straight skeleton, however, need not even be connected.

Lemma 12. There exists a polygon P with holes and an assignment σ of weights to edges such that $\mathcal{S}(P, \sigma)$ is not connected.

Proof. If all weights are negative, then no straight skeleton arcs exist in the interior of P . However, straight skeleton arcs do exist in each component of $\mathbb{R}^2 \setminus P$. Thus, $\mathcal{S}(P, \sigma)$ is not connected.

The following lemma serves as a tool to prove connectedness. The lemma basically says that straight-skeleton features that are connected via the wavefront at any time are also connected within the final straight skeleton.

Lemma 13. Let $\mathcal{S}_t(P, \sigma)$ denote the straight-skeleton features traced by $\mathcal{W}_{P,\sigma}$,σuntil time t. If two points $p, q \in \mathcal{S}_t(P, \sigma)$ are path-connected on $\mathcal{S}_t(P, \sigma) \cup \mathcal{W}_{P,\sigma}(t)$, then they are path-connected on $\mathcal{S}(P, \sigma)$.

Proof. Assume that p and q are two path-connected points on $\mathcal{S}_t(P,\sigma) \cup \mathcal{W}_{P,\sigma}(t)$. We prove this lemma by induction on the events of $\mathcal{W}_{P,\sigma}$ in chronological order. We need to check that connectivity between p and q is maintaineDespite the changes of $\mathcal{W}_{P,\sigma}$ caused by events.

An event may simply remove a collapsed edge of WP,σ(t)WP,σ(t) (edge event), remove a collapsed component of WP,σ(t)WP,σ(t) (edge event), split a component of WP,σ(t)WP,σ(t) (split event), or merge components (split event). In any case, a straight-skeleton node v is created to which arcs are incident that were traced by the vertices of each involved component of the wavefront. However, even if the wavefront is split into multiple components, each component remains connected to v by at least one arc that is traced by a vertex of each component. That is, an event never disconnects a component from the node that is created by the event. 4

By induction, p and q are thus path-connected within $\mathcal{S}_{t'}(P,\sigma) \cup \mathcal{W}_{P,\sigma}(t')$ for t′ after the last event. All remaining wavefront components move towards infinity. As we add an infinite node for each escaping wavefront component, $\mathcal{S}(P,\sigma)$ can be considered to result from $\mathcal{S}_{t'}(P,\sigma) \cup \mathcal{W}_{P,\sigma}(t')$ by contracting each component of $\mathcal{W}_{P,\sigma}(t')$ to a node. Hence, p and q are also path-connected within $\mathcal{S}(P,\sigma)$.

Corollary 14. $\mathcal{S}(P,\sigma)$ is connected for simple polygons.

Proof. This holds even for negative weights as the initial wavefront of P consists of only one connected component.

Lemma 15. $\mathcal{S}(P,\sigma)$ is connected for polygons with holes P and positive weights.

Proof. Consider the wavefront emanating from some hole. At some point, it either merges with some other wavefront during a split event; then the claim holds by induction (and Corollary 14) since WP,σWP,σ then has fewer components. Or it collapses during an edge event. But this is impossible, since for positive weights the wavefront of the hole moves towards the inside of P and hence encloses increasingly more area.

5.4. Bounds on the number of arcs

It is well-known that the unweighted straight skeleton of a simple polygon is a tree [1]. As we will see, for positive weights the weighted straight skeleton is also a tree. We show an even stronger statement, which bounds the number of arcs even in the presence of holes.

In the following, we will use that $P \cup T(P,\sigma)$ is a 2-manifold in order to apply the formula of Euler–Poincaré. As we mentioned at the beginning of Section 5, every event is either an edge event or a split event that was caused by a wavefront vertex meeting the interior of a wavefront edge. It is easy to see that a local neighborhood at each of the corresponding vertices of T(P,σ)T(P,σ) is topologically equivalent to the plane. (**Figure**. 6 shows an example where the corresponding vertex causes T(P,σ)T(P,σ) not to be a 2-manifold. Note that this example does not meet our requirements as two events coincide.) Also recall that we insert an infinite straight-skeleton node for every wavefront polygon that propagates towards infinity. Topologically, we can interpret this as follows: For a sufficiently large time such that all events have already been processed, we topologically glue together the points of C for every remaining wavefront polygon C. Hence, a neighborhood of infinite vertices is again topologically equivalent to the plane.

Lemma 16. Let P be a polygon with n vertices and h holes and let σ be an assignment of arbitrary weights to the edges of P. Assume the body enclosed $P \cup T(P,\sigma)$ has genus g. Denoting the number of straight-skeleton nodes

$$e = n + v - 1 + 2g - h.$$

and arcs by v and e, it holds that

Proof. We consider the body enclosed by $P \cup T(P,\sigma)$. Recall that all faces of $T(P,\sigma)$ are weakly-simple polygons since they have been traced out by edges of the wavefront over time and we add ghost vertices whenever fragments of the same wavefront edge would re-merge. P has, by definition, h holes. We transform P into a weakly-simple polygon P′P′ by adding h edges between pairs of vertices of P. Each of these edges connects one hole of P to the outer boundary of P, either directly or indirectly via already connected holes. The body $P' \cup T(P,\sigma)$ now has $v' = v + n$ vertices, $e' = e + n + h$ edges, and $f' = n + 1$ faces. We can triangulate each (weakly-simple) face without changing the Euler–Poincaré characteristic. Thus, by applying the formula of Euler–Poincaré to the triangulated surface, we obtain $e = n + v - 1 + 2g - h.$

Corollary 17. Let P be a polygon with h holes, and let σ be an assignment of positive weights to the edges of P. Then $e = n + v - 1 + h,$ where v and e denote the number of straight-skeleton nodes and arcs.

Proof. By Lemma 4, $\mathcal{T}(P, \sigma)$ is a terrain that projects to P. Hence the genus g of the body enclosed by $P \cup \mathcal{T}(P, \sigma)$ is equal to h. Thus, Lemma 16 implies that $e = n + v - 1 + h.$

Lemma 18. Let P be a simple polygon. Then $\mathcal{S}(P, \sigma)$ is a tree if and only if the body enclosed by $P \cup \mathcal{T}(P, \sigma)$ has genus zero.

Proof. Let v and e denote the number of straight-skeleton nodes and arcs and let g denote the genus of the body enclosed by $P \cup \mathcal{T}(P, \sigma).$ Since P is simple, $\mathcal{S}(P, \sigma)$ is connected by Corollary 14. S(P,σ)S(P,σ) is a tree if and only if $e = n + v - 1.$ Lemma 16 says that $e = n + v - 1 + 2g.$ Thus S(P,σ)S(P,σ) is a tree if and only if g=0g=0.

Corollary 19. Let P be a simple polygon and let $(e) > 0$ for all wavefront edges e. Then $\mathcal{S}(P, \sigma)$ is a tree.

5.5. Convex polygons

In the following section we investigate the weighted straight skeleton and roof-model of a strictly convex polygon P, i.e., a polygon whose vertices have interior angles less than π.

Lemma 20. There exists a strictly convex polygon P with weights chosen from the set $\{-3, -1, 3\}$ such that $\mathcal{S}(P, \sigma)$ has crossings.

Proof. The polygon is shown in **Figure**. 11.

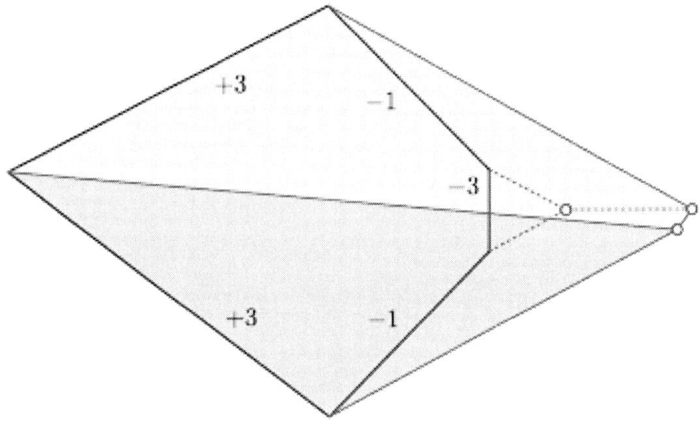

Figure 11. Even if P is a convex polygon, S(P,σ)S(P,σ) may have crossings. The two left-most edges have weight +3, the vertical edge has weight −3, the two remaining edges have weight −1.

Lemma 21. Let P be a strictly convex polygon, and let σ be an assignment of arbitrary weights to its edges. Then the wavefront WP,σ(t)WP,σ(t)is strictly convex at all times $t \geq 0.$

Proof. We first note that the angles between neighboring edges of the wavefront stay constant between events. The initial wavefront is strictly convex as P is convex. By induction over all events (i) every event is an edge event and (ii) the wavefront remains strictly convex after each event.
The following lemma is given as Lemma 6.1 in [13].

Lemma 22. A polyhedron is convex if and only if its edges are convex.

Lemma 23. A polyhedron PPis convex if and only if there is a plane H such that

(i) all intersections of PPwith planes parallel to H are convex, and
(ii)all edges of PPparallel to H are convex.

Proof. If PP is convex then (i) and (ii) are trivial. For the other direction, by Lemma 22 it suffices to show that those edges e that are not parallel to H are convex. We choose a point p in the interior of e and a parallel plane H′ of H that contains p . The intersection $H' \cap P$ gives a convex polygon by (i) and p is a convex vertex of $H' \cap P$ Thus, the edge e is convex.

Theorem 24. Let P be a strictly convex polygon, and let σ be an assignment of arbitrary weights to the edges. Then the body enclosed by $P \cup T(P,\sigma)$ is a convex polyhedron.

Proof. Lemma 21 tells us that the wavefront of a strictly convex polygon is always strictly convex too. Thus, Lemma 23 (i) trivially holds for H being the xy-plane and it remains to check Lemma 23 (ii). Assume that e is an edge in the roof that is horizontal, i.e., parallel to the xy-plane. As the wavefront forms a strictly convex polygon at all times, the horizontal roof edge e must have been created when two parallel wavefront edges meet and, thereby, cause the strictly convex wavefront polygon to collapse. In particular, this implies that e is a ridge and, hence, is convex in the roof, satisfying Lemma 23 (ii). Thus, $P \cup \mathcal{T}(P,\sigma)$ is a convex polyhedron by Lemma 23.

Every straight-skeleton face corresponds to a face of the roof. By Theorem 24, the straight-skeleton faces are not just monotone but in fact convex:

Corollary 25. Let P be a strictly convex polygon, and let σ be an assignment of arbitrary weights to its edges. Then the straight-skeleton faces are convex.

We can even give a characterization of the body $\mathcal{B}$ enclosed by the roof model. Each face of $\mathcal{B}$ is either a face of the roof model (and hence uniquely corresponds to a face $f(e)$ of the straight-skeleton and hence an edge e of P), or it is P. For an edge e of P, let $H_\sigma(e) \subset \mathbb{R}^3$ be the half-space that supports the face of $\mathcal{T}(P,\sigma)$ that corresponds to f(e)f(e) and that has $\mathcal{B}$ inside. Let $H(P) := \mathbb{R}^2 \times [0,\infty)$; this half-space supports P and looks "upward", hence also contains BB. Using E(P) to denote the edges of P, Theorem 24 then implies:

Corollary 26. Let P be a strictly convex polygon, and let σ be an assignment of arbitrary weights to the edges. Then the body BBenclosed by PUT(P,σ)PUT(P,σ)satisfies

$$\mathcal{B} = H(P) \cap \bigcap_{e \in E(P)} H_\sigma(e).$$

A long standing problem of ordinary straight skeletons was to find non-procedural characterizations [3] and [14]. Corollary 26 allows us to characterize any weighted straight skeleton of a convex polygon as the projection of all edges of the polyhedron $H(P) \cap \bigcap_{e \in E(P)} H_\sigma(e)$ to the plane $\mathbb{R}^2 \times \{0\}$. Note that $H_\sigma(e)$ can easily be determined from e and $\sigma(e)$. Since all half-spaces (except for *H(P))* correspond to edges of P and, thus, occur in sorted circular order along P, the results of Aggarwal et al. [15] on computing certain kinds of convex hulls in three dimensions are applicable, thus establishing the proof of the following theorem.

Theorem 27. Let P be a strictly convex n-gon, and let σ be an assignment of arbitrary weights to its edges. Then $S(P,\sigma)$ can be computed in $O(n)$ time.

We also want to remark that the restriction to $H(P)=\mathbb{R}^2\times[0,\infty)$ is only a technicality in the following sense: If we were to extend the definition of $S(P,\sigma)$ and $T(P,\sigma)$ to negative times, then $T(P,\sigma)=\bigcap_{e\in E(P)}H_\sigma(e)$.

Lemma 28. Let P be a strictly convex polygon, and let σ be an assignment of arbitrary weights to the edges. Then S(P,σ)is a tree.

Proof. By Theorem 24, the roof has genus zero. By Lemma 18, S(P,σ) is a tree. (Recall that we insert an infinite straight-skeleton node for the wavefront polygon should it propagate towards infinity.)

Corollary 29. Let P be a strictly convex polygon. For any weight function σ, the straight skeleton S(P,σ) has n+v−1 arcs, where n is the number of vertices of P and v is the number of nodes in S(P,σ).

6. CONCLUSION

In this paper, we studied weighted straight skeletons. We showed that the natural definition already comes with ambiguities and that many properties that seemed to be natural for unweighted straight skeletons do not remain valid for weighted straight skeletons. Hence, caution must be used when applying weighted straight skeletons. A general observation is that if weights are positive or the input polygon is convex then most, albeit not all, properties of straight skeletons carry over to the weighted counterpart.

In some applications of weighted straight skeletons, the weights are chosen in a specific way. In Haunert and Sester [6] the weights are positive and, as we showed in this paper, the weighted straight skeleton is a crossing-free tree, as required by the application. In Barequet et al. [5], however, weighted straight skeletons of polygons are considered, where negative weights in general cannot be avoided.5 Still the weights are chosen in some specific way such that the weighted straight skeleton may have the desired properties. Showing this remains as an open problem.

For convex polygons we were able to characterize the roof and, thus, the straight skeleton as an intersection of half-spaces. Cheng and Vigneron [16] and Huber and Held [14] used the motorcycle graph and the generalized motorcycle graph, respectively, to achieve a similar characterization for ordinary straight skeletons by means of the lower envelope of partially defined linear functions. Generalizing this idea to weighted straight skeletons of non-convex polygons or polygons with holes is an important task and remains open. Similarly, Cheng and Vigneron [16] and Huber and Held [14] used the (generalized) motorcycle graph to compute the straight skeleton of non-degenerate polygons and arbitrary planar straight-line graphs. Can these algorithms be generalized to weighted straight skeletons? This open problem is directly related to characterizing weighted straight skeletons by means of lower envelopes.

In a recent paper [17], we showed how to characterize polygonal tessellations S of the plane to ascertain which form the straight skeleton of some planar straight-line graph G and how to determine G from its putative straight skeleton S. Can those methods be generalized to weighted straight skeletons?

In this paper, we discussed the weighted straight skeleton of polygons and polygons with holes. A generalization of straight skeletons to a non-intersecting collection of polygons is obvious. One may also generalize the weighted straight skeleton to planar straight-line graphs G. For the ordinary straight skeleton, every edge of G sends out a wavefront edge on either side and every terminal vertex of G sends out its own wavefront edge. Hence, it may make sense to restrict the weights to be positive, but the weights of wavefront edges emanating from the same edge of G need not be equal.

Aichholzer et al. [18] used a generalization of straight skeletons where wavefront edges do not necessarily start to propagate at time zero. Such straight skeletons could be called additively-weighted straight skeletons [3], in contrast to multiplicatively-weighted straight skeletons, which are studied in this paper. Systematically investigating additively-weighted straight skeletons remains as future work.

ACKNOWLEDGEMENT

We would like to thank David Eppstein for helpful discussions regarding the ambiguity caused at split events.

REFERENCES

1. O. Aichholzer, D. Alberts, F. Aurenhammer, B. Gärtner, Straight skeletons of simple polygons, in: Proc. 4th Internat. Symp. of LIESMARS, Wuhan, P.R.China, 1995, pp.114–124.
2. O. Aichholzer, F. Aurenhammer, Straight skeletons for general polygonal **Figure**ures in the plane, in: A. Samoilenko (Ed.), Voronoi's Impact on Modern Science, Book 2, Institute of Mathematics of the National Academy of Sciences of Ukraine, Kiev, Ukraine, 1998, pp.7–21.

3. S. Huber, Computing Straight Skeletons and Motorcycle Graphs: Theory and Practice, Shaker Verlag, ISBN978-3-8440-0938-5, 2012.

4. D. Eppstein, J. Erickson, Raising roofs, crashing cycles, and playing pool: applications of a data structure for finding pairwise interactions, Discrete Comput. Geom. 22(4) (1999) 569–592.

5. G. Barequet, D. Eppstein, M.T. Goodrich, A. Vaxman, Straight skeletons of three-dimensional polyhedra, in: Proc. 16th Annu. Europ. Symp. Algorithms, Karlsruhe, Germany, 2008, pp.148–160.

6. J.-H. Haunert, M. Sester, Area collapse and road centerlines based on straight skeletons, GeoInformatica 12 (2008) 169–191.

7. R. Laycock, A. Day, Automatically generating large urban environments based on the footprint data of buildings, in: Proc. 8th Symp. Solid Modeling Applications, Seattle, WA, USA, 2003, pp.346–351.

8. T. Kelly, P. Wonka, Interactive architectural modeling with procedural extrusions, ACM Trans. Graph. 30(2) (2011) 14:1–14:15, http://dx.doi.org/10.1145/1944846.1944854.

9. F. Aurenhammer, Weighted skeletons and fixed-share decomposition, Comput. Geom. Theory Appl. 40(2) (2008) 93–101, http://dx.doi.org/10.1016/j.comgeo.2007.08.002.

10. K. Sugihara, Design of pop-up cards based on weighted straight skeletons, in: Proc. 10th Int. Symp. Voronoi Diagrams in Science & Engineering (ISVD'13), 2013, pp.23–28.

11. T. Kelly, http://code.google.com/p/campskeleton/.

12. T. Biedl, S. Huber, P. Palfrader, Stable roommates for weighted straight skeletons, in: Proc. 30th Europ. Workshop Comput. Geom., Ein Gedi, Israel, 2014.

13. S. Devadoss, J. O'Rourke, Discrete and Computational Geometry, Princeton University Press, ISBN978-0691145532, 2011.

14. S. Huber, M. Held, A fast straight-skeleton algorithm based on generalized motorcycle graphs, Int. J. Comput. Geom. Appl. 22(5) (2012) 471–498, http://dx.doi.org/10.1142/S0218195912500124.

15. A. Aggarwal, L.J. Guibas, J. Saxe, P.W. Shor, A linear-time algorithm for computing the Voronoi diagram of a convex polygon, Discrete Comput. Geom. 4(6) (1989) 591–604.

16. S.-W. Cheng, A. Vigneron, Motorcycle graphs and straight skeletons, Algorithmica 47 (2007) 159–182,

17. T. Biedl, M. Held, S. Huber, Recognizing straight skeletons and Voronoi diagrams and reconstructing their input, in: Proc. 10th Int. Symp. Voronoi Diagrams in Science & Engineering, 2013, pp.37–46.

18. O. Aichholzer, F. Aurenhammer, B. Palop, Quickest paths, straight skeletons, and the city Voronoi diagram, Discrete Comput. Geom. 31(1) (2004) 17–35.

CHAPTER 4

A Quantitative Computational Method for Landslip Orientation of a Body Isolated from Bed Rock in Mountain by Using Spatial Analytic Geometry

Chenggui Huang

Mathematical College of Tianjin Normal University, Upāsaka (Viewer of Reality of Personality), Tianjin Dabei Buddhist Institute, Tianjin, China

ABSTRACT

The paper is going to give a quantitative computational method for "Landslip Orientation of a body isolated from bed rock in Mountain" only with spatial analytic geometry. The paper gives computational formulae in proper order for only landslip plane, just two landslip planes and just three landslip planes, and gives numerical examples. And the paper gives a general computational model for landslip orientation of m landslip planes. The author puts forward "Un-allowed Straight Angle Law", "Critical Straight Angle Law" and "Allowed Straight Angle Law". Finally, the author gives a project expression of a landslip plane on unit sphere.

Keywords: Landslip Plane; Normalization of Horizontal Project; Un-Allowed Straight Angle Law; Critical Straight Angle Law; Allowed Straight Angle Law; Project Expression According to Unit Sphere

1. INTRODUCTION

In paper [1], Professor Shi Gen Hua found that the landslip of an isolated rock body is only able to produce along finite planes. If one parallel moves the finite planes passed through origin O, then he will find on the unite sphere centered O they form a polygon constructed by their up sides. Let S denote the lowest point (or points) of the polygon, if the perpendicular height of point S is lower than the equatorial plane of the sphere, then OS denotes the landslip orientation. In this paper, in order to convenient to engineers, the author provides mainly a quantitative computational method to solve OS. Therefore we suppose that $\Pi_1, \Pi_2, \cdots,$ and Π_n are n possible landslip planes, where n is a natural number. Their equations are as follows1:

$$\Pi_1 : z = a_1 x + b_1 y, \Pi_2 : z = a_2 x + b_2 y, \cdots,$$
$$\text{and } \Pi_n : z = a_n x + b_n y \tag{1}$$

For convenient and united, we let x axis point to east, y axis to north, and z axis perpendicularly to up orientation.

Professor Shi originally used stereographic projection2 to transfer the planes onto the equatorial plane, and he applied topologic method to prove the existence of the orientation of landslip. Furthermore, Shi prevented landslip by piling to withstand landslip along the steepest decent orientation.

2. TO ANALYZE THE LANDSLIP ORIENTATION IN THE CASE OF ONLY ONE LANDSLIP PLANE

The basic supposition anDeduction of compute formulas for the Case of Only One Landslip Plane are as follows.

Here suppose there is only one landslip plane P, whose equation is: $z = ax + by$, where $(a,b) \neq (0,0)$. Its normal is $\boldsymbol{n} = (a,b,-1)$. And the intersection line L of P and the contour plane z = constant C satisfies $ax + by = c$. The normal of $z = c$ is $\boldsymbol{n}_c = (0,0,1)$. Let the orientation of intersection L be $\mathbf{1}$ then it holds

$$\boldsymbol{l} = \boldsymbol{n} \times \boldsymbol{n}_c = \det \begin{pmatrix} \boldsymbol{i} & \boldsymbol{j} & \boldsymbol{k} \\ a & b & -1 \\ 0 & 0 & 1 \end{pmatrix} = (b, a, 0) \tag{2}$$

Let $\boldsymbol{s}$ denote the steepest ascent orientation, then it holds

$$s = l \times n = \det\begin{pmatrix} i & j & k \\ b & -a & 0 \\ a & b & -1 \end{pmatrix} = \left(a, b, a^2 + b^2\right) \tag{3}$$

Let $\rho = \sqrt{a^2 + b^2}$, $a = \rho\cos\theta$, $b = \rho\sin\theta$, then the steepest ascent orientation after normalization of horizontal project is denoted by $s_0 = \frac{s}{\rho}$, then s_0 is expressed as follows

$$s_0 = (\cos\theta, \sin\theta, \rho),\quad \cos\theta = \frac{a}{\rho},\quad \sin\theta = \frac{b}{\rho} \tag{4}$$

where q denotes the horizontal angle in the xy plane during the steepest ascent, r the ascent gradient. From (a,b)one may compute $\mathrm{actg}\frac{b}{a}$. For the signs of (a,b), we have the following eight cases:

$$(+,0),(+,+),(0,+),(-,+),(-,0),(-,-),(0,-),(+,-) \tag{5}$$

If signs are $(+,-)$, then $\theta = \mathrm{actg}\frac{b}{a}$, in the fourth quadrant. Let angle q denote the steepest ascent orientation. The steepest decent angle is $\theta^T = \theta \pm 180^\circ = \theta \pm \pi$, satisfies $0 \le \theta, \theta^T \le 2\pi$. The following open interval

$$\left(\theta - 90^\circ, \theta + 90^\circ\right) = \left(\theta - \frac{\pi}{2}, \theta + \frac{\pi}{2}\right) \tag{6}$$

denotes the angle interval on horizontal plane, in which does not allow slip, called un-allowed interval. However the open interval

$$\left(\theta^T - 90^\circ, \theta^T + 90^\circ\right) = \left(\theta^T - \frac{\pi}{2}, \theta^T + \frac{\pi}{2}\right) \tag{7}$$

denotes the angle interval on horizontal plane, in which allows slip, called allowed interval.

We explain basic properties for the case of Only One Landslip Plane.

Because the normal of the decent slip plane is $n = (a, b, -1)$, and the steepest ascent orientation is $s = \left(a, b, a^2 + b^2\right)$, the horizontal projects of n and s are equal to each other, i.e. equal to (a,b). The orientation of intersectional line of the contour plane z = C and P is $\boldsymbol{l}=(b,-a,0)$, whose horizontal project orientation is (b, -a), which is perpendicular to horizontal project of $\boldsymbol{s}$.

From (a,b) one obtains $\mathrm{actg}\frac{b}{a}$. We have supposed that q is an angle in {Oxy} horizontal plane for the steepest ascent case. If the signs of (a,b) are (+,+), then $\theta=\mathrm{actg}\frac{b}{a}$, in the first quadrant; if the signs of (a,b) are (-,-), $\theta=\pi+\mathrm{actg}\frac{b}{a}$, in the third quadrant. If the signs of (a,b) are (-,+), then $\theta=\pi+\mathrm{actg}\frac{b}{a}$, in the second quadrant. If the signs of (a,b) are (+,-), then $\theta=\mathrm{actg}\frac{b}{a}$, in the fourth quadrant.

The steepest decent angle is $\theta^T=\theta\pm180^\circ=\theta\pm\pi$. In the horizontal plane the un-allowed open interval is $\Delta=\left(\theta-\frac{\pi}{2},\theta+\frac{\pi}{2}\right)$. The allowed interval is $\Delta^T=\left(\theta^T-\frac{\pi}{2},\theta^T+\frac{\pi}{2}\right)$. The intersectional line L of them is their boundary line.

Because the line L is horizontal, the weight is not able to move the isolated rock. However, if there is wind force, water force or earth force to move it, possibly it is able to produce horizontal movement. So the line L is critical line.

We have known that the normal of the slip plane P is $\boldsymbol{n}=(a,b,-1)$, whose horizontal project is (a,b). The horizontal project orientation of intersectional line L is (b,-a), which is the critical orientation between Δ and Δ^T. The horizontal project of P is resolved to be $\Delta\cup L\cup\Delta^T$.

However, the slip orientation $\boldsymbol{s}_0$ after normalization of horizontal project had not been mentioned in Shi's paper [1].

The first case of plane division is as follows. When $a>0$ and $b>0$, q is in the first quadrant of xy plane, the decent orientation from north-east points to southwest. The support must from south-west points to northeast. If a denotes the angle of gradient, then we have

$$\mathrm{tg}\alpha=\rho\text{, and }\alpha=\mathrm{actg}\rho. \tag{8}$$

where ρ denotes the falling gradient of up support, a denotes the angle of altitude.

The included angle between plane P and horizontal plane $z=0$ is equal to a. The intersection line between plane P and horizontal $z=0$ is $ax+by=0$.

Especially, when $a=b=1$, the steepest ascent angle $\theta=45^{\circ}=\frac{\pi}{4}$, the steepest decent angle $\theta^{T}=\theta\pm180^{\circ}=\theta\pm\pi=225^{\circ}=\frac{5\pi}{4}$, and $\rho=\sqrt{2}$, the angle of gradient $\alpha=\text{actg}\sqrt{2}\approx54^{\circ}45'$. The allowed angle interval is $(135^{\circ},315^{\circ})=\left(\frac{3\pi}{4},\frac{7\pi}{4}\right)$. And the angle interval $(-45^{\circ},135^{\circ})=\left(-\frac{\pi}{4},\frac{3\pi}{4}\right)$ is un-allowed to slip.

Now $\theta=45^{\circ}$, the slip orientation points from northeast to south-west. In the practice, at the risk area, the support orientation should point from south-west to north-east according to the horizontal angle q, and to the angle of gradient a by piling to withstand landslip.

3. TO ANALYZE THE OTHER SEVEN CASES FOR THE PLANE DIVISION IN THE CASE OF ONLY ONE LANDSLIP PLANE

The second case of plane division is as follows. When

$a=0$ and $b>0$, $\theta=90^{\circ}=\frac{\pi}{2}$, $\theta^{T}=\frac{\pi}{2}+\pi=\frac{3\pi}{2}$, the slip orientation points from north to south. The support orientation should point from south to north. If a denotes the angle of gradient, then $\text{tg}\alpha=\rho=b$, and $\alpha=\text{actg}\rho$. r is ascent gradient, a is ascent angle. The un-allowed angle interval is

$$\left(\theta-\frac{\pi}{2},\theta+\frac{\pi}{2}\right)=(0,\pi) \tag{9}$$

The allowed angle interval is

$$\left(\theta^{T}-\frac{\pi}{2},\theta^{T}+\frac{\pi}{2}\right)=(\pi,2\pi) \tag{10}$$

The third case of plane division is as follows. When

$a<0$ and $b>0$, $\frac{\pi}{2}<\theta<\pi$, in the second quadrant: $\frac{3\pi}{2}<\theta^T<2\pi$, the slip orientation is from north-west to south-east. The support orientation is from south-east to north-west. If a denotes the angle of gradient, then

$\operatorname{tg}\alpha=\rho=\sqrt{a^2+b^2}$, and $\alpha=\operatorname{actg}\rho$, where r is the gradient of ascent support, a is ascent angle.

The fourth case of plane division is as follows. When $a<0$ and $b=0$, $\theta=\pi$, at -x axis; $\theta^T=0$. The slip orientation is from west to east. The support orientation is from east to west. If a denotes the angle of gradient, then $\operatorname{tg}\alpha=\rho=|a|$, and $\alpha=\operatorname{actg}\rho$, where r is ascent support gradient, a is ascent angle. The un-allowed interval is $\left(\theta-\frac{\pi}{2},\theta+\frac{\pi}{2}\right)=\left(\frac{\pi}{2},\frac{3\pi}{2}\right)$; the allowed interval is $\left(\theta^T-\frac{\pi}{2},\theta^T+\frac{\pi}{2}\right)=\left(-\frac{\pi}{2},+\frac{\pi}{2}\right)$.

The fifth case of plane division is as follows. When

$a<0$ and $b<0$, $\pi<\theta<\frac{3\pi}{2}$, in the third quadrant; $0<\theta^T<\frac{\pi}{2}$, the slip orientation is from south-west to north-east. The support orientation should be from north-east to south-west. If a denotes the angle of gradient, then $\operatorname{tg}\alpha=\rho=\sqrt{a^2+b^2}$, and $\alpha=\operatorname{actg}\rho$, where r is ascent support gradient, a is ascent angle.

The sixth case of plane division is as follows. When

$a=0$ and $b<0$, $\theta=\frac{3\pi}{2}$, at -y axis; $\theta^T=\frac{\pi}{2}$. The slip orientation is from south to north. The support orientation should be from north to south. If a denotes the angle of gradient, then $\operatorname{tg}\alpha=\rho=|b|$, and $\alpha=\operatorname{actg}\rho$, where r is ascent support gradient, a is ascent angle. The un-allowed interval is $\left(\theta-\frac{\pi}{2},\theta+\frac{\pi}{2}\right)=(\pi,2\pi)$; the allowed interval is $\left(\theta^T-\frac{\pi}{2},\theta^T+\frac{\pi}{2}\right)=(0,\pi)$.

The seventh case of plane division is as follows. When $a>0$ and $b<0$, $\frac{3\pi}{2}<\theta<2\pi$, it is in the fourth quadrant, and the slip orientation is from south-east to north-west. The support orientation should be from north-west to south-east. If a denotes the angle of gradient, then $\mathrm{tg}\alpha = \rho = \sqrt{a^2+b^2}$, and $\alpha = \mathrm{actg}\rho$, where r is ascent support gradient, a is ascent angle.

The eighth case of plane division is as follows. When $a > 0$ and $b = 0$, $\theta = 0^\circ$, at +x axis, and $\theta^T = \pi$. The slip orientation is from east to west. The support orientation should be from west to east. If a denotes the angle of gradient, then $\mathrm{tg}\alpha = \rho = a$, and $\alpha = \mathrm{actg}\rho$, r denotes the ascent gradient of supporting, a is the angle of ascent supporting. The un-allowed interval is $\left(\theta - \frac{\pi}{2}, \theta + \frac{\pi}{2}\right) = \left(-\frac{\pi}{2}, +\frac{\pi}{2}\right)$; the allowed interval is $\left(\theta^T - \frac{\pi}{2}, \theta^T + \frac{\pi}{2}\right) = \left(\frac{\pi}{2}, \frac{3\pi}{2}\right)$.

4. THE CASE OF JUST HAVING TWO SLIP PLANES

Now we suppose that Π_1 and Π_2 are two possible slip planes. If Π_1 and Π_2 are parallel each other, and because the two planes have a common point: (0,0,0), they must be coincident, and turn to be the case of only one slip plane, as stated in the former section. Therefore we only discuss the case of that the two slip planes Π_1 and Π_2 have an intersectional line. Suppose the equations of the two slip planes are as follows;

Suppose $\Pi_1 : z = a_1 x + b_1 y$, $\Pi_2 : z = a_2 x + b_2 y$. The two planes have a common point: (0,0,0). Because we have supposed that Π_1 and Π_2 have an intersectional line $L_{1,2}$, then $a_1 = a_2$, and $b_1 = b_2$ are not able to hold simultaneously. Because of the normal of Π_1 is $\boldsymbol{n}_1 = (a_1, b_1, -1)$, and the normal of Π_2 is $\boldsymbol{n}_2 = (a_2, b_2, -1)$. It follows that the orientation $\boldsymbol{l}_{1,2}$ of $L_{1,2}$ is denoted by3

$$\boldsymbol{l}_{1,2} = \boldsymbol{n}_1 \times \boldsymbol{n}_2 = \det\begin{pmatrix} \boldsymbol{i} & \boldsymbol{j} & \boldsymbol{k} \\ a_1 & b_1 & -1 \\ a_2 & b_2 & -1 \end{pmatrix}$$

$$= (-b_1 + b_2, -a_2 + a_1, a_1 b_2 - a_2 b_1) \tag{11}$$

Let $\rho_{1,2} = \sqrt{(b_2 - b_1)^2 + (a_1 - a_2)^2}$, after normalized horizontal project we have

$$\frac{l_{1,2}}{\rho_{1,2}} = \left(\cos\theta_{1,2}, \sin\theta_{1,2}, h_{1,2}\right) \tag{12}$$

where

$$\cos\theta_{1,2} = \frac{(b_2 - b_1)}{\rho_{1,2}}, \sin\theta_{1,2} = \frac{(a_1 - a_2)}{\rho_{1,2}},$$

$$\text{and } h_{1,2} = \frac{(a_1 b_2 - a_2 b_1)}{\rho_{1,2}} \tag{13}$$

Let $l_{1,2,0}$ denote the slip orientation under normalizetion horizontal project, then it holds the following formula

$$l_{1,2,0} = -\operatorname{sgn} h_{1,2}\left(\cos\theta_{1,2}, \sin\theta_{1,2}, h_{1,2}\right) \tag{14}$$

Or in detail, when $h_{1,2} > 0$, it holds

$$l_{1,2,0} = -\left(\cos\theta_{1,2}, \sin\theta_{1,2}, h_{1,2}\right) \quad , (15)$$

while the slip horizontal angle is $\theta^T_{1,2} = \theta_{1,2} \pm \pi$.

When $h_{1,2} = 0$, it holds

$$l_{1,2,0} = \pm\left(\cos\theta_{1,2}, \sin\theta_{1,2}, 0\right) \quad , (16)$$

while it is not able to slip, called critical case.

When $h_{1,2} < 0$, it holds

$$l_{1,2,0} = \left(\cos\theta_{1,2}, \sin\theta_{1,2}, 0\right) \quad , (17)$$

while the slip horizontal angle is $\theta^T_{1,2} = \theta_{1,2}$

And the absolute value $|h_{1,2}| = h_{1,2} \times \operatorname{sgn} h_{1.2}$ denotes the decent gradient of intersection line $L_{1,2}$. Besides if $a_{1,2}$ denotes the ascent angle for slip, then it holds

$$\operatorname{tg} a_{1,2} = |h_{1,2}|, \ \operatorname{actg}|h_{1,2}| = a_{1,2} \tag{18}$$

5. EXAMPLE FOR JUST TWO SLIP PLANES

Suppose $\Pi_1 : z = x + y$, $\Pi_2 : z = -x + 3y$. For

$\Pi_1 : a_1 = b_1 = 1$. $\theta_1 = \frac{\pi}{4}$, The un-allowed interval is $\Delta_1 = \left(-\frac{\pi}{4}, \frac{3\pi}{4}\right) = (0.785398, 2.35619)$. And $\rho_1 = \sqrt{2}$, the ascent angle $\alpha_1 = \text{actg}\sqrt{2} \approx 0.955317$, θ_1 is in the first quadrant in xy horizontal plane. The slip orientation is from north-east to south-west. $\theta_1^T = \frac{5\pi}{4}$ is the steepest decent horizontal angle. The support orientation should be from south-west to north-east. The allowed interval is $\Delta_1^T = \left(\frac{3\pi}{4}, \frac{7\pi}{4}\right) = (2.35619, 5.49779)$.

For $\Pi_2 : a_2 = -1, b_2 = 3$. The signs are (-,+), $\rho_1 = \sqrt{10}$, $90^\circ < \theta_2 < 180^\circ$ is in the second quadrant, $\cos\theta_2 = -\frac{\sqrt{10}}{10}$, $\sin\theta_2 = \frac{3\sqrt{10}}{10}$, $\text{tg}\theta_2 = -3$, the slip orientation is from north-west to south-east. The support orientation is from north-west to south-east. $\theta_2 = \text{actg}(-3) = 1.89254$. The un-allowed interval is $\Delta_2 = (0.321751, 3.46334)$; the allowed interval is $\Delta_2^T = (-2.81985, 0.321751)$.

The ascent angle $\alpha_2 = \text{actg}\sqrt{10} \approx 1.26452$.

The intersection of two allowed interval is as follows

$$\Delta_{1,2}^T = \Delta_1^T \cap \Delta_2^T = (3.46334, 5.49779) \tag{19}$$

We have supposed that the intersectional line of Π_1 and Π_2 is $L_{1,2}$, where $b_2 - b_1 = 2$, $a_1 - a_2 = 2$, it follows

$$\boldsymbol{l}_{1,2} = \boldsymbol{n}_1 \times \boldsymbol{n}_2 = \det\begin{pmatrix} \boldsymbol{i} & \boldsymbol{j} & \boldsymbol{k} \\ a_1 & b_1 & -1 \\ a_2 & b_2 & -1 \end{pmatrix} = \begin{pmatrix} \boldsymbol{i} & \boldsymbol{j} & \boldsymbol{k} \\ 1 & 1 & -1 \\ -1 & 3 & -1 \end{pmatrix} = (2, 2, 4) \tag{20}$$

$$\rho_{1,2} = \sqrt{(b_2 - b_1)^2 + (a_1 - a_2)^2} = 2\sqrt{2} \tag{21}$$

$$\cos\theta_{1,2} = \frac{(b_2 - b_1)}{\rho_{1,2}} = \frac{\sqrt{2}}{2} \tag{22}$$

$$\sin\theta_{1,2} = \frac{(a_1 - a_2)}{\rho_{1,2}} = \frac{\sqrt{2}}{2} \tag{23}$$

$$\mathrm{tg}\theta_{1,2} = 1; \quad \theta_{1,2} = 45^\circ = \frac{\pi}{4} = 0.78539815 \tag{24}$$

And the decent gradient

$$h_{1,2} = \frac{(a_1 b_2 - a_2 b_1)}{\rho_{1,2}} = \sqrt{2} > 0 \tag{25}$$

Refer to Equation (15), because $h_{1,2} > 0$, it holds:

$$l_{1,2,0} = -(\cos\theta_{1,2}, \sin\theta_{1,2}, h_{1,2}) \tag{26}$$

while the steepest slip horizontal angle is

$$\theta^T_{1,2} = \theta_{1,2} \pm \pi = \frac{5\pi}{4} = 3.92699075$$

Very fortunately, here the slip orientation of $L_{1,2}$ is: $l_{1,2,0} = -(\cos\theta_{1,2}, \sin\theta_{1,2}, h_{1,2})$, which coincides with $-s_0$ when $a = b = 1$, by using the formula in Equation (4) according to condition $a = b = 1$. Note that here Π_1 is namely the plane Π when $a = b = 1$. The steepest decent angle is $\theta^T_{1,2} = \theta_{1,2} \pm \pi = 225^\circ = \frac{5\pi}{4} = 3.92699075$ and $\rho_1 = \sqrt{2}$, the ascent angle is $\alpha_{1,2} = \mathrm{actg}\sqrt{2} \approx 0.955317 \approx 54^\circ 45'$. The allowed interval is $(135^\circ, 315^\circ) = \left(\frac{3\pi}{4}, \frac{7\pi}{4}\right)$. And the un-allowed interval is $(-45^\circ, 135^\circ) = \left(-\frac{\pi}{4}, \frac{3\pi}{4}\right)$. $\theta^T_{1,2} = 3.92699075 \in \Delta^T_1 \cap \Delta^T_2 = (3.46334, 5.49779)$.

Very fortunately, here we have $\theta^T_{1,2} = \theta^T_1 = \frac{5\pi}{4}$. In dealing with the landslip of an isolated rock in a tunnel, we should note not only the caving in along the slip orientation $-s_0$ of the intersectional line $L_{1,2}$ of Π_1 and Π_2; but also note the landslip along the steepest decent orientation $-s_0$ of the plane Π_1. Therefore, the support area should be some what wider, while we shoulDecide according to practical survey.

Note the following data for comparing:

$$\frac{\pi}{4}=0.78539815;\ \frac{\pi}{2}=1.5707963;$$

$$\frac{3\pi}{4}=2.35619445;\ \pi=3.1415926;$$

$$\frac{5\pi}{4}=3.92699075;\ \frac{3\pi}{2}=4.7123889;$$

$$\frac{7\pi}{4}=5.49778705;\ 2\pi=6.2831825;$$

To sum up, the damage orientations are three:

The slip orientation along the intersectional line $L_{1,2}$

$$l_{1,2,0}=-\left(\cos\theta_{1,2},\sin\theta_{1,2},h_{1,2}\right)=-\left(\cos 45^\circ,\sin 45^\circ,\sqrt{2}\right) \tag{27}$$

The steepest horizontal normalized slip orientation along P1:

$$-s_{1,0}=-\left(\cos\theta_1,\sin\theta_1,\rho_1\right)=-\left(\frac{1}{\sqrt{2}},\frac{1}{\sqrt{2}},\sqrt{2}\right) \tag{28}$$

The steepest horizontal normalized slip orientation along P2;

$$\begin{aligned}-s_{2,0}&=-\left(\cos\theta_2,\sin\theta_2,\rho_2\right)\\&=-\left(-\frac{\sqrt{10}}{10},\frac{3\sqrt{10}}{10},\sqrt{10}\right)=\left(\frac{\sqrt{10}}{10},-\frac{3\sqrt{10}}{10},-\sqrt{10}\right)\end{aligned} \tag{29}$$

Fortunately, it holds here: $l_{1,2,0}=-s_{1,0}$. If having not done through repeated computations, only by intuition, it is difficult to see the result. In the paper [1] of Professor Shi Gen Hua, he claimed that the slip orientation $l_{1,2,0}$ of the intersectional line of the most damage one, the author also had the view. However the slip orientations $-\mathbf{s}_{1,0}$ and $-\mathbf{s}_{2,0}$ had not be mentioned, which should be noted also.

According to the data of the problem, the most damage orientations are three: 1) The slip orientation $\mathbf{l}_{1,2,0}$ along the intersectional line $L_{1,2}$; 2) The steepest slip orientation $-\mathbf{s}_{1,0}$ along Π_1; and 3) The steepest slip orientation $-\mathbf{s}_{2,0}$ along Π_2. Because the existence of Π_1 forms a bound for the orientation $-\mathbf{s}_{2,0}$; but the existence of Π_2 does not form a bound for the orientation $-\mathbf{s}_{1,0}$. That needs a detail analysis for three orientations, please refer to the following minimal analysis.

Because the section appears fortunate case: $\mathbf{l}_{1,2,0} = -\mathbf{s}_{1,0}$, we must note special cases which appear occasionally like this kind.

Checking computations: Substituting $(x, y) = (-1, -1)$ into Π_1 we get: $z = x + y = -2$. Substituting $(x, y) = (-1, -1)$ into Π_2 we get: $z = -x + 3y = -2$. It is certainly the intersection of Π_1 and Π_2.

The following is to find Allowed Slip Pyramid. Refer to **Figure**ure 1, the slip horizontal angle is ÐPOR. The allowed slip three edge pyramid formed by two planes Π_1 and Π_2 is as follows: The first edge is OP. The first boundary surface is Π_2. The second edge is $\mathbf{l}_{1,2}$ which is under the ray q1, 2T. The second boundary surface is Π_1. The third edge is OR. The third boundary surface is the horizontal plane from OR to OP along clockwise.

The following is Minimal Analysis.

The horizontal project **Figure**ure of Π_1 and Π_2.The allowed interval is $\angle POR = \Delta_{1,2}^{T} = (3.46334, 5.49779)$.

Fortunately it holds $\theta_1^T = \theta_{1,2}^T = \frac{5\pi}{4} = 3.92699075$ which is the first minimal angle. However

$\theta_2^T = 5.03413 = 1.60241\pi$, which falls in the fourth quadrant and beyond the scope of ÐPO - q1, 2T, we do not take it. Therefore, in the section, the steepest slip horizontal angle is namely $\theta_1^T = \theta_{1,2}^T = \frac{5\pi}{4} = 3.92699075$. The ascent angle is $\alpha = \text{actg}\sqrt{2} = 0.955317$.

6. EXAMPLE FOR JUST THREE SLIP PLANES

Suppose the first landslip plane Π_1 is: $z = a_1x + b_1y = x + y$. Through computation, the normal of Π_1 is: $\boldsymbol{n}_1 = (a_1, b_1, -1) = (1, 1, -1)$,

$$\theta_1 = \text{actg}\frac{b_1}{a_1} = \text{actg}1 = 0.785398, \quad \theta_1^T = 3.92699.$$

$$\rho_1 = \sqrt{a_1^2 + b_1^2} = \sqrt{2}, \quad \alpha_1 = \text{actg}\rho_1 = 0.955317.$$

$$\boldsymbol{s}_{1,0} = (\cos\theta_1, \sin\theta_1, \rho_1) = \left(\frac{1}{\sqrt{2}}, \frac{1}{\sqrt{2}}, \sqrt{2}\right);$$

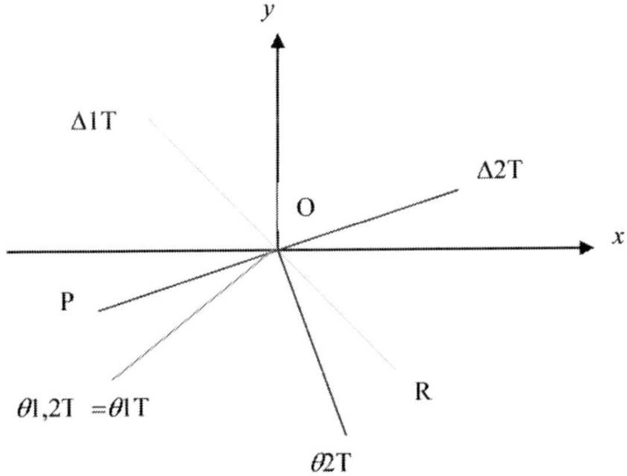

Figure 1. The horizontal project **Figure**ure for $\boldsymbol{\Pi_1}$ and Π_2.

the un-allowed interval is

$$\Delta_1 = \left(\theta_1 - \frac{\pi}{2}, \theta_1 + \frac{\pi}{2}\right) = (-0.785398, 2.35619);$$

and the allowed interval is

$$\Delta_1^T = \left(\theta_1^T - \frac{\pi}{2}, \theta_1^T + \frac{\pi}{2}\right) = (2.35619, 5.4779).$$

Suppose the second landslip plane Π_{10} is: $z = a_{10}x + b_{10}y = 2x + y$; through computation we get: the normal of Π_{10} is:

$$\boldsymbol{n}_{10} = (a_{10}, b_{10}, -1) = (2, 1, -1),$$

$$\theta_{10} = \text{actg}\frac{b_{10}}{a_{10}} = 0.463648,$$

$$\theta_{10}^T = 3.60524.$$

$$\rho_{10} = \sqrt{a_{10}^2 + b_{10}^2} = \sqrt{5},$$

$$\alpha_{10} = \text{actg}\rho_{10} = 1.15026.$$

$$\boldsymbol{s}_{10,0} = (\cos\theta_{10}, \sin\theta_{10}, \rho_{10}) = \left(\frac{2}{\sqrt{5}}, \frac{1}{\sqrt{5}}, \sqrt{5}\right);$$

The un-allowed interval is

$$\Delta_{10} = \left(\theta_{10} - \frac{\pi}{2}, \theta_{10} + \frac{\pi}{2}\right) = (-1.10715, 2.03444);$$

and the allowed interval is

$$\Delta_{10}^T = \left(\theta_{10}^T - \frac{\pi}{2}, \theta_{10}^T + \frac{\pi}{2}\right) = (2.03444, 5.17604).$$

It follows that the orientation $\boldsymbol{l}_{1,10}$ of intersectional line $L_{1,10}$ is denoted by

$$\boldsymbol{l}_{1,10} = \boldsymbol{n}_1 \times \boldsymbol{n}_{10} = \det\begin{pmatrix} \boldsymbol{i} & \boldsymbol{j} & \boldsymbol{k} \\ a_1 & b_1 & -1 \\ a_{10} & b_{10} & -1 \end{pmatrix}$$

$$= \begin{pmatrix} \boldsymbol{i} & \boldsymbol{j} & \boldsymbol{k} \\ 1 & 1 & -1 \\ 2 & 1 & -1 \end{pmatrix} = (0,-1,-1) \tag{30}$$

Let $\rho_{1,10} = \sqrt{(b_{10}-b_1)^2 + (a_1 - a_{10})^2} = 1$, and let $\boldsymbol{l}_{1,10,0}$ denote horizontal normalizeDecent slip orientation, and $\theta_{1,10}$ the decent slip angle in the horizontal project plane, $\cos\theta_{1,10} = 0$, $\sin\theta_{1,10} = -1$, and $h_{1,10} = -1$; And the horizontal signs of $\boldsymbol{l}_{1,10}$ is (0,-), then $\theta_{1,10}$ is on -y axisit holds $\theta_{1,10} = \frac{3\pi}{2} = 4.7123889$, and

$$\boldsymbol{l}_{1,10,0} = (0,-1,-1),\ \theta_{1,10} = 4.7123889 \tag{31}$$

Because of it holds $\theta^T_{1,10} = \theta_{1,10} = 4.7123889$; the decent slip orientation $\boldsymbol{l}_{1,10,0}$ of $L_{1,10}$ is a horizontal project normalized orientation. The angle

$$\theta^T_{1,10} = \theta_{1,10} = 4.7123889 \tag{32}$$

On one hand, we must note the first paragraph in Section 6, the steepest decent slip orientation of Π_1 is

$$-\boldsymbol{s}_{1,0} = -(\cos\theta_1, \sin\theta_1, \rho_1) = -\left(\frac{1}{\sqrt{2}}, \frac{1}{\sqrt{2}}, \sqrt{2}\right)$$

$$= \left(-\frac{1}{\sqrt{2}}, -\frac{1}{\sqrt{2}}, -\sqrt{2}\right) \tag{33}$$

The allowed interval is

$$\Delta^T_1 = (2.35619, 5.49779) \tag{34}$$

On the other hand, through computation we get: the steepest decent slip orientation of P10 is

$$-\boldsymbol{s}_{10,0} = -(\cos\theta_{10}, \sin\theta_{10}, \rho_{10}) = -\left(\frac{2}{\sqrt{5}}, \frac{1}{\sqrt{5}}, \sqrt{5}\right)$$
$$= \left(-\frac{2}{\sqrt{5}}, -\frac{1}{\sqrt{5}}, -\sqrt{5}\right) \quad (35)$$

The allowed interval is

$$\Delta_{10}^{T} = \left(\theta_{10}^{T} - \frac{\pi}{2}, \theta_{10}^{T} + \frac{\pi}{2}\right) = (2.03444, 5.17604) \quad (36)$$

Because of the intersection of Δ_1^T and Δ_{10}^T is

$$\Delta_{1,10}^{T} = (2.35619, 5.17604) \quad (37)$$

This is the alloweDecent slip interval.

Refer to Equation (37), it holds the inequalities

$$2.35619 < 4.7123889 < 5.17604 \quad (38)$$

It namely holds the membership

$$\theta_{1,10}^{T} = \theta_{1,10} = 4.7123889 \in \Delta_{1,10}^{T} = (2.35619, 5.17604) \quad (39)$$

And note that $\alpha_{1,10}$ denotes the ascent angle of the slip, it follows

$$\mathrm{tg}\alpha_{1,10} = |h_{1,10}| = |-1| = 1,\ \alpha_{1,10} = \mathrm{actg}1 = \frac{\pi}{4} = 0.78539815 \quad (40)$$

Suppose the third landslip plane Π_2 is: $z = a_2x + b_2y = -x + 3y$. Through computation we get: the normal of Π_2 is $\boldsymbol{n}_2 = (a_2, b_2, -1) = (-1, 3, -1)$,

$$\theta_2 = \mathrm{actg}\frac{b_2}{a_2} = \pi - 1.24905 = 1.89254,$$

$$\theta_2^T = -1.24905.$$

$$\rho_2 = \sqrt{a_2^2 + b_2^2} = \sqrt{10},\ \alpha_2 = \mathrm{actg}\rho_2 = 1.26452.$$

$$s_{2,0} = (\cos\theta_2, \sin\theta_2, \rho_2) = \left(-\frac{1}{\sqrt{10}}, \frac{3}{\sqrt{10}}, \sqrt{10}\right);$$

The un-allowed interval is

$$\Delta_2 = \left(\theta_2 - \frac{\pi}{2}, \theta_2 + \frac{\pi}{2}\right) = (0.321751, 3.46334);$$

and the allowed interval is

$$\Delta_2^T = \left(\theta_2^T - \frac{\pi}{2}, \theta_2^T + \frac{\pi}{2}\right) = (-2.81985, 0.321751).$$

It follows that the orientation $l_{2,10}$ of intersectional line $L_{2,10}$ is denoted by

$$l_{2,10} = n_2 \times n_{10} = \det\begin{pmatrix} i & j & k \\ a_2 & b_2 & -1 \\ a_{10} & b_{10} & -1 \end{pmatrix}$$

$$= \begin{pmatrix} i & j & k \\ -1 & 3 & -1 \\ 2 & 1 & -1 \end{pmatrix} = (-2, -3, -7) \tag{41}$$

Let $\rho_{2,10} = \sqrt{(b_{10} - b_2)^2 + (a_2 - a_{10})^2} = \sqrt{13}$, and let $l_{2,10,0}$ denote the horizontal normalizeDecent slip orientation, and $\theta_{2,10}$ the slip angle in horizontal plane, $\cos\theta_{2,10} = -\frac{2}{\sqrt{13}}$, $\sin\theta_{2,10} = -\frac{3}{\sqrt{13}}$, and $h_{2,10} = -\frac{7}{\sqrt{13}}$; And $\mathrm{tg}\theta_{2,10} = \frac{3}{2} = 1.5$, $\mathrm{actg}1.5 = 0.98279$ and the horizontal signs of $l_{2,10}$ are (-,-), then $\theta_{2,10}$ is in the third quadrant, it holds $\theta_{2,10} = \mathrm{actg}\frac{3}{2} + \pi = 3.1415926 + 0.98279 = 4.12438$, which is between ($\frac{5\pi}{4} = 3.92699075$; $\frac{3\pi}{2} = 4.7123889$). We have the following formulae

$$\theta_{2,10} = 4.12438 \quad l_{2,10,0} = \left(-\frac{2}{\sqrt{13}}, -\frac{3}{\sqrt{13}}, -\frac{7}{\sqrt{13}}\right) \tag{42}$$

Because it holds $\theta_{2,10}^{T}=\theta_{2,10}=4.12438$; the decent slip orientation $l_{2,10,0}$ of $L_{2,10}$ is horizontal project normalization. The angle

$$\theta_{2,10}^{T}=\theta_{2,10}=4.12438 \tag{43}$$

On one hand, through computation we get: the steepest decent slip orientation along Π_2 is

$$-s_{2,0}=-(\cos\theta_2,\sin\theta_2,\rho_2)=-\left(-\frac{1}{\sqrt{10}},\frac{3}{\sqrt{10}},\sqrt{10}\right)$$
$$=\left(\frac{1}{\sqrt{10}},-\frac{3}{\sqrt{10}},-\sqrt{10}\right) \tag{44}$$

The allowed interval is

$$\Delta_2^{T}=(2.81985,0.321751) \tag{45}$$

On the other hand, through computation we get:, the steepest decent slip orientation along Π_{10} is

$$-s_{10,0}=-(\cos\theta_{10},\sin\theta_{10},\rho_{10})=-\left(\frac{2}{\sqrt{5}},\frac{1}{\sqrt{5}},\sqrt{5}\right)$$
$$=\left(-\frac{2}{\sqrt{5}},-\frac{1}{\sqrt{5}},-\sqrt{5}\right) \tag{46}$$

The allowed interval is

$$\Delta_{10}^{T}=\left(\theta_{10}^{T}-\frac{\pi}{2},\theta_{10}^{T}+\frac{\pi}{2}\right)=(2.03444,5.17604) \tag{47}$$

The intersection of Δ_2^{T} and Δ_{10}^{T} is

$$\Delta_{2,10}^{T}=(3.46334,5.17604) \tag{48}$$

This is the allowed interval of the slip.

Refer to Equation (43), it holds the inequalities

$$3.46334<4.12438<5.17604 \tag{49}$$

It namely holds the membership

$$\theta_{2,10}^{T} = 4.12438 \in \Delta_{2,10}^{T} = (3.46334, 5.17604) \tag{50}$$

And note that $\alpha_{2,10}$ denotes the ascent angle of the slip, it holds

$$\mathrm{tg}\alpha_{2,10} = |h_{2,10}| = \left| -\frac{7}{\sqrt{13}} \right| = \frac{7}{\sqrt{13}} \tag{51}$$

$$\alpha_{2,10} = \mathrm{actg}\frac{7}{\sqrt{13}} = \mathrm{actg}1.941450 = 1.095159 = 0.34851\pi \tag{52}$$

We have supposed that the intersectional line of Π_1 and Π_2 is $L_{1,2}$, where $b_2 - b_1 = 2$, $a_1 - a_2 = 2$, it follows

$$\boldsymbol{l}_{1,2} = \boldsymbol{n}_1 \times \boldsymbol{n}_2 = \det\begin{pmatrix} \boldsymbol{i} & \boldsymbol{j} & \boldsymbol{k} \\ a_1 & b_1 & -1 \\ a_2 & b_2 & -1 \end{pmatrix} = \begin{pmatrix} \boldsymbol{i} & \boldsymbol{j} & \boldsymbol{k} \\ 1 & 1 & -1 \\ -1 & 3 & -1 \end{pmatrix} = (2, 2, 4) \tag{53}$$

$$\rho_{1,2} = \sqrt{(b_2 - b_1)^2 + (a_1 - a_2)^2} = 2\sqrt{2} \tag{54}$$

$$\cos\theta_{1,2} = \frac{(b_2 - b_1)}{\rho_{1,2}} = \frac{\sqrt{2}}{2} \tag{55}$$

$$\sin\theta_{1,2} = \frac{(a_1 - a_2)}{\rho_{1,2}} = \frac{\sqrt{2}}{2} \tag{56}$$

$$\mathrm{tg}\theta_{1,2} = 1;\ \theta_{1,2} = 45^\circ = \frac{\pi}{4} = 0.78539815 \tag{57}$$

And the decent slip gradient is

$$h_{1,2} = \frac{(a_1 b_2 - a_2 b_1)}{\rho_{1,2}} = \sqrt{2} > 0 \tag{58}$$

Refer to Equation (15), because of $h_{1,2} > 0$, it holds

$$\boldsymbol{l}_{1,2,0} = -(\cos\theta_{1,2}, \sin\theta_{1,2}, h_{1,2}) \tag{59}$$

The steepest decent slip horizontal angle is

$$\theta^T_{1,2} = \theta_{1,2} \pm \pi = \frac{5\pi}{4} = 3.92699075$$

Fortunately, here the decent slip orientation of $L_{1,2}$ is $\boldsymbol{l}_{1,2,0} = -(\cos\theta_{1,2}, \sin\theta_{1,2}, h_{1,2})$, which is coincide with $-\boldsymbol{s}_0$ in section 2, above Equation (8). Note that here Π_1 is namely the plane Π in the example in section 2, when $a = b = 1$. The steepest decent angle is

$\theta^T_{1,2} = \theta_{1,2} \pm \pi = 225^\circ = \frac{5\pi}{4} = 3.92699075$, and $\rho = \sqrt{2}$, the ascent angle $\alpha = \text{actg}\sqrt{2} \approx 0.955317 \approx 54^\circ 45'$.

The allowed interval is

$$\Delta^T_1 = (135^\circ, 315^\circ) = \left(\frac{3\pi}{4}, \frac{7\pi}{4}\right) \tag{60}$$

The un-allowed interval is

$$\Delta_1 = (-45^\circ, 135^\circ) = \left(-\frac{\pi}{4}, \frac{3\pi}{4}\right) \tag{61}$$

$$\theta^T_{1,2} = 3.92699075 \in \Delta^T_1 \cap \Delta^T_2 = (3.46334, 5.49779) \tag{62}$$

For three planes Π_1, Π_2 and Π_{10}, the allowed interval is

$$\Delta^T_{1,2,10} = \Delta^T_1 \cap \Delta^T_2 \cap \Delta^T_{10} = (3.46334, 5.17604) \tag{63}$$

Therefore it holds the memberships

$$\theta^T_{1,2} = 3.92699075 \in \Delta^T_{1,2,10} = (3.46334, 5.17604) \tag{64}$$

$$\theta^T_{2,10} = 4.12438 \in \Delta^T_{1,2,10} = (3.46334, 5.17604) \tag{65}$$

$$\theta^T_{1,10} = 4.7123889 \in \Delta^T_{1,2,10} = (3.46334, 5.17604) \tag{66}$$

Correspondingly, the ascent angles are

$$\alpha_{1,2} = \text{actg}\sqrt{2} = 0.955317 = 0.304087\pi \tag{67}$$

$$\alpha_{2,10} = \text{actg}\frac{7}{\sqrt{13}} = 1.095159 = 0.34851\pi \tag{68}$$

$$\alpha_{2,10} = \text{actg}1 = \frac{\pi}{4} = 0.78539815 \tag{69}$$

And it holds the inequalities

$$\pi < \theta_{1,2}^{T} < \theta_{2,10}^{T} < \theta_{1,10}^{T} = \frac{3\pi}{2} \tag{70}$$

$$\frac{\pi}{4} = \alpha_{1,10} < \alpha_{1,2} < \alpha_{2,10} = 1.095159 = 0.34851\pi \tag{71}$$

Along counter clockwise, from Π_{10} to Π_1 to Π_2, now $\alpha_{2,10}$ is the steepest decent gradient. How to analyze? At first we list the main data for planes as follows

$$\Pi_{10} : \theta_{10} = 0.463648;\ \alpha_{10} = 1.15026;\ \Delta_{10}^{T} = (2.03444, 5.17604) \tag{72}$$

$$\Pi_1 : \theta_1 = 0.785398;\ \alpha_1 = 0.955317;\ \Delta_1^{T} = (2.35619, 5.49779) \tag{73}$$

$$\Pi_2 : \theta_2 = 1.89254;\ \alpha_2 = 1.26452;\ \Delta_2^{T} = (3.46334, 6.60494) \tag{74}$$

Afterwards, we list the main data for intersectional lines as follows

$$L_{1,2} : \theta_{1,2}^{T} = 3.92699075;\ \alpha_{1,2} = 0.955317;\ \Delta_{1,2}^{T} = (3.46334, 5.49779) \tag{75}$$

$$L_{2,10} : \theta_{2,10}^{T} = 4.12438;\ \alpha_{2,10} = 1.095159;\ \Delta_{2,10}^{T} = (3.46334, 5.17604) \tag{76}$$

$$L_{1,10} : \theta_{1,10}^{T} = 4.7123889;\ \alpha_{1,10} = \frac{\pi}{4} = 0.78539815;\ \Delta_{1,10}^{T} = (2.35619, 5.17604) \tag{77}$$

Now we draw the orientation **Figure**ure of horizontal project as follows We give explanations for the **Figure**ure 2. Let the coordinates {Oxy} denote the horizontal project plane The straight angle under the line D2T-O-P is denoted by $\Delta_2^{T} = (3.46334, 6.60494)$.

The straight angle on the left of the line D10T-O-P is denoted by $\Delta_{10}^{T} = (2.03444, 5.17604)$.

The straight angle on the left of the line D1T-O-R is denoted by $\Delta_1^T = (2.35619, 5.49779)$.

The angle of ray q1, 10T is denoted by

$$\theta_{1,10}^T = 4.7123889$$.

The angle of ray q2, 10T is denoted by

$$\theta_{2,10}^T = 4.12438$$.

The angle of ray q1, 2T is denoted by

$$\theta_{1,2}^T = \theta_{1,2} \pm \pi = \frac{5\pi}{4} = 3.92699075$$.

The angle ĐPOQ denotes the allowed interval

$$\Delta_{1,2,10}^T = (3.46334, 5.17604)$$

Pyramid Allowed to Decent Slip

We start from OP, and let OP be the first edge. According to counter clockwise, the first boundary surface of pyramid is Π_2. Through computation we have $z_{10} = -1.75$, $z_1 = -1$, and $z_2 = 0$.

Refer to the **Figure**ure 2, according to counter clockwise, the second edge of the pyramid is the edge $l_{1,2}$ which is under the ray q1, 2T. Through computation, according to counter clockwise, the second boundary surface of the pyramid is Π_1. $z_{10} = -1.5$, $z_1 = -1$, $z_2 = -1$.

Refer to the **Figure** 2, according to counter clockwise, the third edge of the pyramid is the edge $L_{2,10}$ which is under the ray q2, 10T. Through computation, according to counter clockwise the third boundary surface is still Π_1. The third edge under q2,10T is covered by Π_1. $z_{10} = -1.4$, $z_1 = -1$, $z_2 = -1.4$.

Refer to the **Figure** 2, according to counter clockwise, the fourth edge of the pyramid is the edge $L_{1,10}$ which is under the ray q1, 10T. Through computation, according to counter clockwise, the fourth boundary surface of the pyramid is Π_{10} or Π_1. $z_{10} = -1$, $z_1 = -1$, $z_2 = -3$.

Refer to the **Figure** 2, according to counter clockwise, the fifth edge of the pyramid is OQ. Through computation, according to counter clockwise, the fourth boundary surface is defined by Π_{10}. $z_{10} = 0$, $z_1 = -0.33333$, $z_2 = -2.33334$.

Refer to the **Figure** 2, the fifth boundary surface of the pyramid is defined by the horizontal plane which is from OQ to OP according to clockwise.

To sum up, the alloweDecent slip pyramiDegenerates to be a four edge pyramid, whose boundary surface begin with the edge OP, through Π_2 to the edge under q1, 2T, and through Π_1 to the edge $L_{1,10}$ under q1, 10T, and through Π_{10} arrived OQ, and according to clockwise, through horizontal plane to the edge OP.

Through minimal computation, the final conclusion is as follows: in the four edge pyramid of the section, $\theta = \theta_1^T = \theta_{1,2}^T = \frac{5\pi}{4}$ is the steepest decent slip orientation.

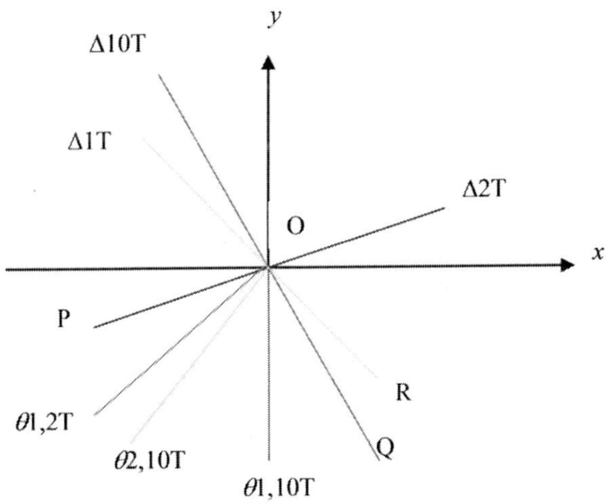

Figure 2. The orientation **Figure** of horizontal project.

We should firstly reinforce along the orientation, i.e. from south-west to north-east, with ascent angle $\alpha = \mathrm{actg}\sqrt{2} = 0.955317 \approx 54^{\circ}45'$, to reinforce.

Minimal value, i.e. the steepest decent slip orientation is along the edge $l_{1,2}$ which is under the ray q1, 2T

7. DISCUSSIONS FOR THE CASE OF M SLIP PLANES

According to Section 1, let $\Pi_1, \Pi_2, \cdots,$ and Π_m be m possible slip planes, where m is a natural number. Their equations are expressed by $\Pi_1 : z = a_1 x + b_1 y, \Pi_2 : z = a_2 x + b_2 y, \cdots,$ and $\Pi_m : z = a_m x + b_m y$. For convenient and unification, the x axis points to east, the y axis points to north, and the z axis points to up.

Suppose $1 \le i \ne j \le m$

Among them, we choose an element Π_i, whose equation is $z = a_i x + b_i y$. Its normal is $\boldsymbol{n}_i = (a_i, b_i, -1)$. And the intersectional line L_i of the contour plane $z = c$ and Π_i satisfies $a_i x + b_i y = c$. The normal of $z = c$ is $\boldsymbol{n}_c = (0,0,1)$. Let $\boldsymbol{l}_i$ be the orientation of intersectional line L_i, it follows

$$\boldsymbol{l}_i = \boldsymbol{n}_i \times \boldsymbol{n}_c = \det \begin{pmatrix} \boldsymbol{i} & \boldsymbol{j} & \boldsymbol{k} \\ a_i & b_i & -1 \\ 0 & 0 & 1 \end{pmatrix} = (b_i, -a_i, 0) \tag{78}$$

Let $\boldsymbol{s}_i$ denote the steepest ascent orientation, it follows

$$\boldsymbol{s}_i = \boldsymbol{l}_i \times \boldsymbol{n}_i = \det \begin{pmatrix} \boldsymbol{i} & \boldsymbol{j} & \boldsymbol{k} \\ b_i & -a_i & 0 \\ a_i & b_i & -1 \end{pmatrix} = (a_i, b_i, a_i^2 + b_i^2) \tag{79}$$

Let $\overline{\boldsymbol{s}_i}$ denote the horizontal project of the steepest ascent orientation $\boldsymbol{s}_i$, which is also the horizontal project of normal $\boldsymbol{n}_i$, i.e. it holds

$$\overline{s_i} = (a_i, b_i) \tag{80}$$

This is a deputation of plane Π_i.

Let $\rho_i = \sqrt{a_i^2 + b_i^2}$, $b_i = \rho_i \sin\theta_i$ $a_i = \rho_i \cos\theta_i$, the horizontal project decent slip orientation is denoted by $s_{i,0} = \frac{s_i}{\rho_i}$, then $s_{i,0}$ is able to be expressed by

$$s_{i,0} = (\cos\theta_i, \sin\theta_i, \rho_i),\ \cos\theta_i = \frac{a_i}{\rho_i},\ \sin\theta_i = \frac{b_i}{\rho_i} \tag{81}$$

Let $\overline{s_{i,0}}$ denote the horizontal normalized project of the steepest ascent orientation, then it follows

$$\overline{s_{i,0}} = (\cos\theta_i, \sin\theta_i) \tag{82}$$

The intersectional line of two planes Π_i and Π_j is $L_{i,j}$, whose orientation $\boldsymbol{l}_{i,j}$ is expressed as follows

$$\begin{aligned}\boldsymbol{l}_{i,j} &= \boldsymbol{n}_i \times \boldsymbol{n}_j = \det\begin{pmatrix} \boldsymbol{i} & \boldsymbol{j} & \boldsymbol{k} \\ a_i & b_i & -1 \\ a_j & b_j & -1 \end{pmatrix} \\ &= (-b_i + b_j, -a_j + a_i, a_i b_j - a_j b_i)\end{aligned} \tag{83}$$

Let $\rho_{i,j} = \sqrt{(b_j - b_i)^2 + (a_i - a_j)^2}$, after horizontal project normalizing one obtains

$$\frac{\boldsymbol{l}_{i,j}}{\rho_{i,j}} = (\cos\theta_{i,j}, \sin\theta_{i,j}, h_{i,j}) \tag{84}$$

where

$$\cos\theta_{i,j} = \frac{(b_j - b_i)}{\rho_{i,j}},\ \sin\theta_{i,j} = \frac{(a_i - a_j)}{\rho_{i,j}},$$
$$\text{and } h_{i,j} = \frac{(a_i b_j - a_j b_i)}{\rho_{i,j}} \tag{85}$$

Let $\boldsymbol{l}_{i,j,0}$ denote the decent slip orientation of horizontal project normalization, then it holds

$$\boldsymbol{l}_{i,j,0} = -\operatorname{sgn} h_{i,j} (\cos\theta_{i,j}, \sin\theta_{i,j}, h_{i,j}) \tag{86}$$

Or in detail, when $h_{i,j} > 0$, it holds:

$$\boldsymbol{l}_{i,j,0} = -\left(\cos\theta_{i,j}, \sin\theta_{i,j}, h_{i,j}\right)$$, (87)

while the decent slip horizontal angle is $\theta_{i,j}^{T} = \theta_{i,j} \pm \pi$

When $h_{i,j} = 0$, it holds

$$\boldsymbol{l}_{i,j,0} = \pm\left(\cos\theta_{i,j}, \sin\theta_{i,j}, 0\right)$$, (88)

while don't produce slip, it may produce horizontal movement, it is critical line.

When $h_{i,j} < 0$, it holds

$$\boldsymbol{l}_{i,j,0} = -\left(\cos\theta_{i,j}, \sin\theta_{i,j}, h_{i,j}\right) \quad (89)$$

while the decent slip horizontal angle is $\theta_{i,j}^{T} = \theta_{i,j}$

And the absolute value $\left|h_{i,j}\right| = h_{i,j} \times \operatorname{sgn} h_{i,j}$ denotes the decent slip gradient of the intersectional line $L_{i,j}$. And if $a_{i,j}$ denotes the ascent angle of the decent slip, then it holds

$$\operatorname{tg}a_{i,j} = \left|h_{i,j}\right|, \quad a_{i,j} = \operatorname{actg}\left|h_{i,j}\right| \quad (90)$$

The steepest decent slip horizontal angle is

$\theta_i^{T} = \theta_i \pm 180^{\circ} = \theta_i \pm \pi$. The un-allowed horizontal angle open interval is $\Delta_i = \left(\theta_i - \frac{\pi}{2}, \theta_i + \frac{\pi}{2}\right)$. The allowed interval is $\Delta_i^{T} = \left(\theta_i^{T} - \frac{\pi}{2}, \theta_i^{T} + \frac{\pi}{2}\right)$. The horizontal intersectional line L_i is their boundary line.

7.1. "Un-Allowed Straight Angle Law" about m Planes

Now we draw the **Figure** of the un-allowed straight angle law about m planes as follows.

In the **Figure** 3, we draw the m steepest ascent angle $\theta_1, \cdots, \theta_i, \cdots, \theta_j, \cdots$, and θ_m whose corresponding rays are denoted by $\theta_1, \cdots, \theta_i, \cdots, \theta_j, \cdots$, and θ_m and the corresponding m planes are $\Pi_1,\cdots,\Pi_i,\cdots,\Pi_j,\cdots$, and Π_m. Suppose they permute according to counter clockwise.

There is an angle between two adjacent rays. Suppose all the angles are less than p, i.e. less than a straight angle, then the un-allowed intervals $\Delta_1,\cdots,\Delta_i,\cdots,\Delta_j,\cdots$, and Δ_m formed by m planes $\Pi_1,\cdots,\Pi_i,\cdots,\Pi_j,\cdots$, and Π_m cover the whole circumference angle. Therefore, there is not any horizontal angle which is alloweDecent slip. This may be simply called un-allowed straight angle law.

7.2. "Critical Straight Angle Law" about m Planes

Now we draw the **Figure** of the critical straight angle law about m planes as follows.

In the **Figure** 4, we draw the m steepest ascent angles $\theta_1,\cdots,\theta_i,\cdots,\theta_j,\cdots$, and θ_m whose corresponding rays are denoted by $\theta_1,\cdots,\theta_i,\cdots,\theta_j,\cdots$, and θ_m and the corresponding m planes are $\Pi_1,\cdots,\Pi_i,\cdots,\Pi_j,\cdots$, and Π_m. Suppose they permute according to counter clockwise.

There is an angle between two adjacent rays. Suppose among the angles there is an angle equal to p, i.e. equal to a straight angle. No harm of the generality, suppose the angle between qm and q1 is equal to p, then the critical line $L_{1,m}$ between Π_m and Π_1 is namely the critical line among $\Pi_1,\cdots,\Pi_i,\cdots,\Pi_j,\cdots$, and Π_m.

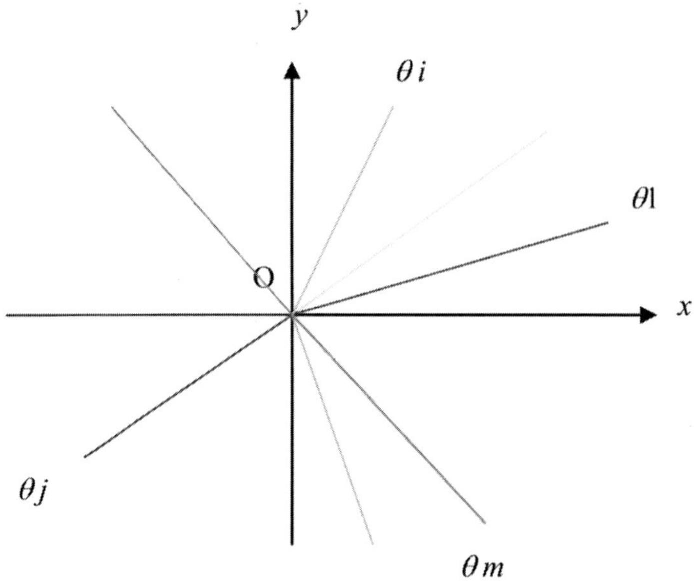

Figureure 3. The **Figure**ure of un-allowed straight angle law.

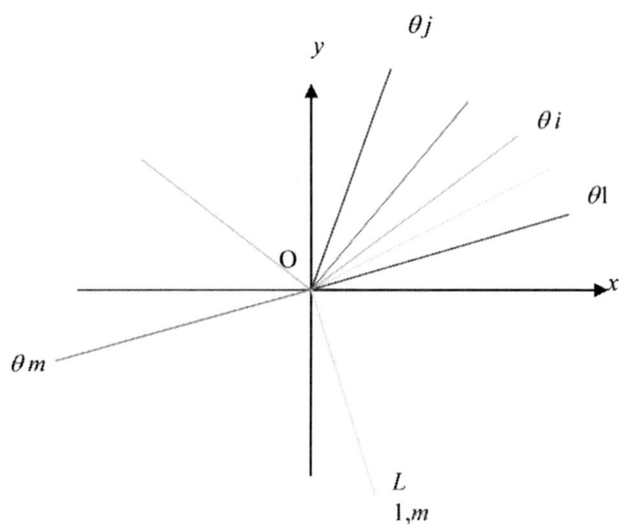

Figureure 4. The **Figure**ure for critical straight angle law.

This is called critical straight angle law. If there is wind force, water force or earth force to move it, possibly it is able to produce horizontal movement.

7.3. "Allowed Straight Angle Law" about m Planes

Now we draw the **Figure**ure of the allowed straight angle law about m planes as follows.

In the **Figure** 5, we draw the m steepest ascent angle $\theta_1,\cdots,\theta_i,\cdots,\theta_j,\cdots,$ and θ_m whose corresponding rays are denoted by $\theta_1,\cdots,\theta_i,\cdots,\theta_j,\cdots,$ and θ_m and the corresponding m planes are $\Pi_1,\cdots,\Pi_i,\cdots,\Pi_j,\cdots,$ and Π_m. Suppose they permute according to counter clockwise.

There is an angle between two adjacent rays. No harm of generality, suppose along counter clockwise the angle ÐqmOq1 is more than p, i.e. more than a straight angle. The angle ÐqmOP is equal to a right angle, then the angle ÐPOQ which is more than zero, is the intersectional part of all allowed angles $\Delta_1,\cdots,\Delta_i,\cdots,\Delta_j,\cdots,$ and Δ_m, and is called allowed interval D. Because of the existence of allowed interval D, which is called allowed straight angle law.

Let $L_{1,m}$ denote the intersectional line of Π_1 and Π_m Generally let $L_{i,j}$ denote the intersectional line of planes Π_i and Π_j, then the set $\{L_{i,j}\}$ have $\frac{m(m-1)}{2}$ elements. And there are m plane steepest decent slip angles $\theta_1^T,\cdots,\theta_i^T,\cdots,\theta_j^T,\cdots,$ and θ_m^T, which is denoted by the set $\{\theta_i^T\}$.

Let the intersectional set $\{L_{i,j}\}\cap\Delta$ denote the set of all intersectional lines in the angle ÐPOQ. Let the intersectional set $\{\theta_i^T\}\cap\Delta$ denote the set of all steepest decent slip angles in the angle ÐPOQ. Afterwards, we

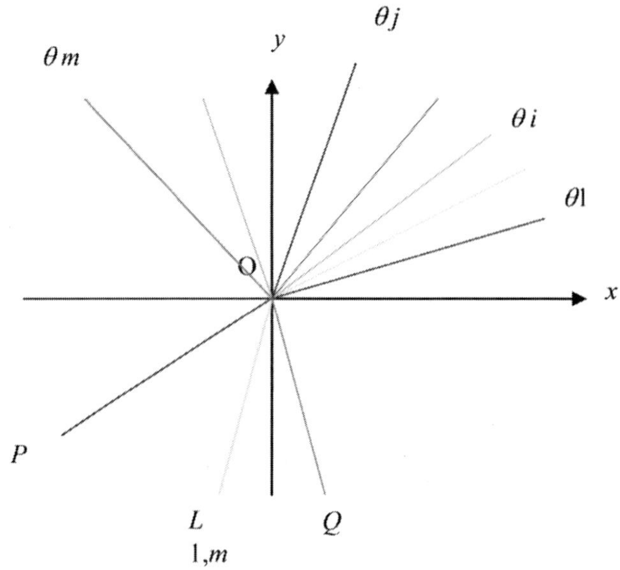

Figureure 5. The **Figure**ure of allowed straight angle law.

compare the z values along these orientations on unit sphere. How to analytically compare? Please refer to Section 6: the example for just have three decent slip planes, in which we should give up some planes and orientations. After sifting we need only analytically compare the values on the left orientations, in which the orientation of the minimal z value is just the steepest decent slip angle formed by m planes $\Pi_1,\cdots,\Pi_i,\cdots,\Pi_j,\cdots,$ and Π_m. This is the most important orientation which need reinforce.

8. PROJECT EXPRESSION OF LANDSLIP PLANE ON UNIT SPHERE

If thinking of the theoretical beauty-ness, one should naturally use the unit spherical project

8.1. Inspiration from the Computation along One Orientation

Now let $(a,b)=(1,2)$ to be an example, while the allowed interval is

$$\Delta^T=\left(\theta^T-\frac{\pi}{2},\theta^T+\frac{\pi}{2}\right)=(2.67795,5.81954)$$
,

$\theta = 1.10715$, $\theta^T = 4.24875$. The ascent angle $\alpha = \text{actg}\rho = 1.15026$. $\cos\alpha = 0.40825$, $\sin\alpha = 0.91287$.

The component $z = -\sin\alpha = -0.91287$, is the vertical component of the intersectional point of the ray formed by (θ^T, α) and on the lower semi unit sphere. Because of $\rho = \sqrt{a^2 + b^2} = \sqrt{5}$, the steepest ascent orientation is $(\cos\theta, \sin\theta, \rho) = \left(\frac{1}{\sqrt{5}}, \frac{2}{\sqrt{5}}, \sqrt{5}\right)$, and the steepest decent slip orientation is

$$\left(\cos\theta^T, \sin\theta^T, -\rho\right) = \left(-\frac{1}{\sqrt{5}}, -\frac{2}{\sqrt{5}}, -\sqrt{5}\right).$$

The equation of unit sphere is $x^2 + y^2 + z^2 = 1$. Let $\sqrt{x^2 + y^2} = r$, $r = |\cos\alpha|$. And let

$$x = r\cos\theta^T = |\cos\alpha|\cos\theta^T,$$
$$y = r\sin\theta^T = |\cos\alpha|\sin\theta^T \qquad (91)$$

Now $\cos\alpha = 0.40825$, it follows

$$x = 0.40825 \times \cos 4.24875 = -0.18257;$$
$$y = 0.40825 \times \sin 4.24875 = -0.36515. \qquad (92)$$

8.2. An Example of Practical Computation on the Project of Unit Sphere

We consider on plane z = 0 the following ellipse whose long radius is equal to 1, and short radius is equal to $|\cos\alpha|$. The orientation of long radius is x, pointed to $\theta^T + \frac{\pi}{2}$. The orientation of short radius is h, pointed to q.

The equation of the ellipse in the coordinates {xOh} is as follows

$$\xi^2 + \frac{\eta^2}{|\cos\alpha|^2} = 1 \qquad (93)$$

We plan to solve the equation of the ellipse in the coordinates {xOy}. In order to doing this, we should find the corresponding formula of the coordinate transformation. Therefore we should firstly draw the hint **Figure** of coordinate transformation.

The hint **Figure** 6 for coordinate transformation is as follows.

In the **Figure** y denotes the angle from Ox axis to Ox axis, whose measure is equal to the angle from $\theta^T + \frac{\pi}{2}$ to Ox axis. In the problem, refer to the first and second paragraphs of article 8.1, because of $\mathrm{tg}\theta = \frac{b}{a}$, we have

$$\psi = \mathrm{actg}\frac{a}{b} = \mathrm{actg}\frac{1}{2} = 0.46365 = 2\pi - \left(\theta^T + \frac{\pi}{2}\right) \\ = 2\pi - 5.81954 \tag{94}$$

The deduction of the formula for coordinate transformation: let

$$\xi = \xi + \mathrm{i}\eta, \quad w = x + \mathrm{i}y \tag{95}$$

Because of

$$\zeta = we^{\mathrm{i}\psi} = w(\cos\psi + \mathrm{i}\sin\psi) \tag{96}$$

It follows the formula

$$\zeta + \mathrm{i}\eta = (x + \mathrm{i}y)(\cos\psi + \mathrm{i}\sin\psi) \\ = x\cos\psi - y\sin\psi + \mathrm{i}(x\sin\psi + y\cos\psi) \tag{97}$$

Finally, we obtain

$$\zeta = x\cos\psi - y\sin\psi; \quad \eta = x\sin\psi + y\cos\psi \tag{98}$$

The above elliptic Equation (93) changes into

$$(x\cos\psi - y\sin\psi)^2 + \frac{(x\sin\psi + y\cos\psi)^2}{|\cos\alpha|^2} = 1 \tag{99}$$

For the decent slip plane P, the allowed interval is $\Delta^T = \left(\theta^T - \frac{\pi}{2}, \theta^T + \frac{\pi}{2}\right) = (2.67795, 5.81954)$, $\theta^T = 4.24874$. Now we divide Δ^T into equidistance 10^n components. In practice the n should be 2, 3, or 4. In the section being only a hint, so we take $n = 1$. Note $\frac{\pi}{10} = +0.31415926$. Similarly, we have the data as follows Let

$\varphi_0 = 2.67795$, $\varphi_1 = \varphi_0 + 0.31415926 = 2.99211$, $\varphi_2 = \varphi_1 + 0.31415926 = 3.30627$,

$\varphi_3 = \varphi_2 + 0.31415926 = 3.62043$,

$\varphi_4 = \varphi_3 + 0.31415926 = 3.93459$,

$\varphi_5 = \varphi_4 + 0.31415926 = 4.24875$,

$\varphi_6 = \varphi_5 + 0.31415926 = 4.56291$,

$\varphi_7 = \varphi_6 + 0.31415926 = 4.87707$,

$\varphi_8 = \varphi_7 + 0.31415926 = 5.19123$,

$\varphi_9 = \varphi_8 + 0.31415926 = 5.50539$,

$\varphi_{10} = \varphi_9 + 0.31415926 = 5.81995$

In the plane $z=0$ for the ray with orientation j, its orientation vector is $(\cos\varphi, \sin\varphi)$, its normal is $(-\sin\varphi, \cos\varphi)$, and its equation is as follows

$$-x\sin\varphi + y\cos\varphi = 0 \tag{100}$$

Note $\psi = 0.46365$, $\cos\psi = 0.89443$,

$\sin\psi = 0.44722$. And note $\cos\alpha = 0.40825$,

$|\cos\alpha|^2 = 0.40825^2 = 0.16667$. The equation of the ellipse is as follows

$$(x\times 0.89443 - y\times 0.44722)^2 + \frac{(x\times 0.44722 + y\times 0.89443)^2}{|0.40825|^2} = 1 \tag{101}$$

$$(x\times 0.89443 - y\times 0.44722)^2 + (x\times 0.44722 + y\times 0.89443)^2 \div |0.40825|^2 = 1 \tag{102}$$

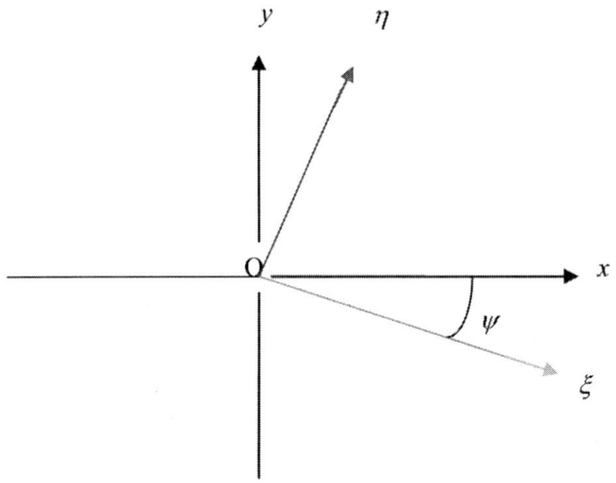

Figure 6. The hint **Figure** for coordinate transformation formula.

The detail numerical computation is as follows :

For $\varphi_0 = 2.67795$, its orientation vector is $(\cos\varphi_0, \sin\varphi_0) = (-0.89443, 0.44721)$. Because its signs are (-,+), the solution (x,y) falls in the second quadrant. Then we use the normal equation: $-x\sin\varphi + y\cos\varphi = -x \times 0.44721 - y \times 0.89443 = 0$. Substituting into the ellipse in Equation (102) one obtains:

$$(x \times 0.89443 - y \times 0.44721)^2 = 1;$$

$$x \times 0.89443 - y \times 0.44721 = \pm 1.$$

When it holds: $x \times 0.89443 - y \times 0.44721 = 1$, from the normal equation one obtains:

$y = \dfrac{-x \times 0.44721}{0.89443} = -0.49999 \times x \approx -\dfrac{1}{2}x$. Substituting into $x \times 0.89443 - y \times 0.44721 = 1$; $0.89443 \times \left(x + \dfrac{1}{4}x\right) = 1$; one obtains $x = \dfrac{1}{0.89443} \times \dfrac{4}{5} > 0$, which does not fall in the second quadrant, so we give up it.

When it holds: $x\times 0.89443-y\times 0.44721=-1$, still from the normal equation one obtains: $y=-\frac{1}{2}x$. Substituting it into $x\times 0.89443-y\times 0.44721=-1$, $0.89443\times\left(x+\frac{1}{4}x\right)=-1$, one obtains: $x=\frac{1}{0.89443}\times\frac{4}{5}=-0.89442$; $y=-\frac{1}{2}x=0.44721$. It is just wanted, the computation is correct. $z_0=\sqrt{1-x^2-y^2}=\sqrt{0.00002}\approx 0$, coinciding.

For $\varphi_1=2.99211$, firstly we compute the orientation vector: $(\cos\varphi_1,\sin\varphi_1)=(-0.98885,0.14893)$. Because its signs are (-,+), the solution (x,y) falls into the second quadrant. Then we use the normal equation: $-x\sin\varphi+y\cos\varphi=-x\times 0.14893-y\times 0.9885=0$; $y=-0.15061x$ Substituting into the ellipse in Equation (102) one obtains:

$$(x\times 0.89443+0.15061x\times 0.44722)^2+(x\times 0.44722-0.15061x\times 0.89443)^2\div|0.40825|^2=1$$

$$x^2\left\{0.96179^2+0.31251^2\div\left|0.40825\right|^2\right\}=1;x^2\left\{0.92504+0.58597\right\}=1;1.22923x=\pm 1$$

By taking the negative, one obtains: $x=-0.81352$; $y=0.12252$. $z_1=-\sqrt{1-x^2-y^2}=-0.56848$.

For $\varphi_2=3.30627$, firstly, we compute the orientation vector: $(\cos\varphi_2,\sin\varphi_2)=(-0.98647,-0.16393)$. Because its signs are (-,-), the solution falls in the third quadrant. Then by using the normal equation one obtains: $-x\sin\varphi+y\cos\varphi=x\times 0.16393-y\times 0.0.98647=0$;

$y=0.16618x$. Substituting into the ellipse in Equation (102), one obtains

$$(x\times 0.89443-0.16618x\times 0.44722)^2$$

$$+(x\times 0.44722+0.16618x\times 0.89443)^2\div|0.40825|^2=1$$

$$x^2\left\{0.82011^2+0.59586^2\div\left|0.40825\right|^2\right\}=1;$$

$$x^2\left\{0.67258+2.13028\right\}=1;1.67417x=\pm 1$$

By taking the negative, one obtains:

$$x = -0.59731;\ y = -0.09926.$$

$$z_1 = -\sqrt{1-x^2-y^2} = -0.79584.$$

For $\varphi_3 = 3.62043$, firstly, we compute the orientation vector: $(\cos\varphi_3, \sin\varphi_3) = (-0.88753, -0.46075)$. Because its signs are (-,-), the solution (x,y) falls in the third quadrant. Then by using the normal equation one obtains: $-x\sin\varphi + y\cos\varphi = x \times 0.46.75 - y \times 0.88753 = 0$; $y = 0.51914x$. Substituting into the ellipse in Equation (102) one obtains

$$(x \times 0.89443 - 0.51914x \times 0.44722)^2 + (x \times 0.44722 + 0.51914x \times 0.89443)^2 \div |0.40825|^2 = 1$$

$$x^2\left\{0.66226^2 + 0.0.91155^2 \div |0.40825|^2\right\} = 1;\ x^2\{0.43859 + 4.9855\} = 1;\ 2.32897x = \pm 1$$

By taking the negative, one obtains:

$$x = -0.42937;\ y = -0.2229.$$

$$z_3 = -\sqrt{1-x^2-y^2} = -0.87519.$$

For $\varphi_4 = 3.93459$, firstly, we compute the orientation vector: $(\cos\varphi_4, \sin\varphi_4) = (-0.70171, -0.71246)$. Because its signs are (-,-), the solution (x,y) falls in the third quadrant. Then by using the normal equation one obtains: $-x\sin\varphi + y\cos\varphi = x \times 0.71246 - y \times 0.70171 = 0$; $y = 1.01532x$. Substituting into the ellipse in Equation (102) one obtains:

$$(x \times 0.89443 - 1.01532x \times 0.44722)^2 + (x \times 0.44722 + 1.01532x \times 0.89443)^2 \div |0.40825|^2 = 1$$

$$x^2\left\{0.44036^2 + 1.35535^2 \div |0.40825|^2\right\} = 1;\ x^2\{0.19392 + 11.02175\} = 1;\ 3.34898x = \pm 1$$

.

By taking the negative one obtains:

$$x = -0.2986;\ y = -0.30317.$$

$$z_3 = -\sqrt{1-x^2-y^2} = -0.90495.$$

For $\varphi_5 = 4.24875$, firstly, we compute the orientation vector: $(\cos\varphi_5, \sin\varphi_5) = (-0.44721, -0.89443)$. Because its signs are (-,-), the solution (x,y) falls in the third quadrant. Then by using the normal equation one obtains: $-x\sin\varphi + y\cos\varphi = x \times 0.89443 - y \times 0.44721 = 0$; $y = 2.00002x$. Substituting into the ellipse in Equation (102) one obtains:

$$\left(x\times0.89443-2x\times0.44722\right)^2+\left(x\times0.44722+2x\times0.89443\right)^2\div\left|0.40825\right|^2=1$$

$$x^2\left\{0^2+2.23608^2\div\left|0.40825\right|^2\right\}=1;\ x^2\left\{30.00007\right\}=1;\ 5.47723x=\pm1$$.

By taking the negative, one obtains: $x_5=-0.18257$; $y_5=-0.36514$.

$z_5=-\sqrt{1-x^2-y^2}=-0.91288$, that coincides with

$\sin\alpha_1=0.91287$ in the first paragraph in Article 8.1. Therefore the computation is correct.

For $\varphi_6=4.56291$, firstly, we compute the orientation vector: $\left(\cos\varphi_6,\sin\varphi_6\right)=\left(-0.14892,-0.98885\right)$. Because its signs are (-,-), the solution (x,y) falls in the third quadrant. Then by using the normal equation one obtains: $-x\sin\varphi+y\cos\varphi=x\times0.98885-y\times0.14892=0$;

$y=6.64014x$. Substituting into the ellipse in Equation (102) one obtains:

$$\left(x\times0.89443-6.64014x\times0.44722\right)^2+\left(x\times0.44722+6.64014x\times0.89443\right)^2\div\left|0.40825\right|^2=1$$

$$x^2\left\{2.07517^2+6.38636^2\div\left|0.40825\right|^2\right\}=1;\ x^2\left\{249.01785\right\}=1;\ 15.7803x=\pm1$$

By taking the negative, one obtains: $x=-0.06337$; $y=-0.42079$. $z_6=-\sqrt{1-x^2-y^2}=-0.90494$. It coincides with $z=-0.90495$ of $\varphi_4=3.93459$. So the computation is reasonable.

For $\varphi_7=4.87707$, firstly, we compute the orientation vector: $\left(\cos\varphi_7,\sin\varphi_7\right)=\left(0.16394,-0.98647\right)$ Because the signs are (+,-), the solution (x,y) falls in the fourth quadrant. Then by using the normal equation one obtains: $-x\sin\varphi+y\cos\varphi=-x\times0.98647-y\times0.16394=0$;

$y=-6.01726x$. Substituting into the ellipse in Equation (102) one obtains:

$$\left(x\times0.89443+6.01726x\times0.44722\right)^2+\left(x\times0.44722-6.01726x\times0.89443\right)^2\div\left|0.40825\right|^2=1$$

$$x^2\left\{3.58547^2+4.9348^2\div\left|0.40825\right|^2\right\}=1;\ x^2\left\{158.96788\right\}=1;\ 12.60825x=\pm1$$.

By taking the positive, one obtains: $x = 0.07931$; $y = -0.47723$. $z_7 = -\sqrt{1-x^2-y^2} = -0.87519$ It coincides with $z = -0.87519$ of $\varphi_3 = 3.62043$. Therefore, the computation is reasonable.

For $\varphi_8 = 5.19123$, firstly, we compute the orientation vector: $(\cos\varphi_8, \sin\varphi_8) = (0.46075, -0.88753)$. Because the signs are (+,-), the solution (x,y) falls in the fourth quadrant. Then by using the normal equation one obtains: $-x\sin\varphi + y\cos\varphi = x\times 0.88753 + y\times 0.46075 = 0$; $y = -1.92627x$. Substituting into the ellipse in Equation (102) one obtains:

$$(x\times 0.89443 + 1.92627x\times 0.44722)^2 + (x\times 0.44722 - 1.92627x\times 0.89443)^2 \div |0.40825|^2 = 1$$

$$x^2\left\{1.7559^2 + 1.27569^2 \div |0.40825|^2\right\} = 1;\ x^2\{12.84743\} = 1;\ 3.58433x = \pm 1$$

.

By taking the positive, one obtains: $x = 0.27899$; $y = -0.53741$. $z_8 = -\sqrt{1-x^2-y^2} = -0.79584$. It coincides with $z = -0.79584$ of $\varphi_2 = 3.30627$. Therefore, the computation is reasonable.

For $\varphi_9 = 5.50539$, firstly, we compute the orientation vector: $(\cos\varphi_9, \sin\varphi_9) = (0.71246, -0.70171)$. Because the signs are (+,-), the solution (x,y) falls in the fourth quadrant. Then by using the normal equation one obtains: $-x\sin\varphi + y\cos\varphi = +x\times 0.70171 + y\times 0.71246 = 0$; $y = -0.98491x$. Substituting into the ellipse in Equation (102) one obtains:

$$(x\times 0.89443 + 0.98491x\times 0.44722)^2 + (x\times 0.44722 - 0.98491x\times 0.89443)^2 \div |0.40825|^2 = 1$$

$$x^2\left\{1.3349^2 + 0.43371^2 \div |0.40825|^2\right\} = 1;\ x^2\{2.91057\} = 1;\ 1.70604x = \pm 1$$

.

By taking the positive, one obtains: $x = 0.58615$;

$y = -0.5773$. $z_9 = -\sqrt{1-x^2-y^2} = -0.56847$. It coincides with $z = -0.56848$ of $\varphi_2 = 2.99211$. Therefore, the computation is reasonable.

For $\varphi_{10} = 5.81955$, firstly, we compute the orientation vector $(\cos\varphi_{10}, \sin\varphi_{10}) = (0.89443, -0.4472)$. Because the signs are (+,-), the solution (x,y) falls in the fourth quadrant. Then by using the normal equation one obtains: $-x\sin\varphi + y\cos\varphi = +x\times 0.4472 + y\times 0.89443 = 0$;

$y = -0.49998x = -\frac{1}{2}x$. Substituting into the ellipse in Equation (102) one obtains:

$$(x\times 0.89443 + 0.5x\times 0.44722)^2 + (x\times 0.44722 - 0.5x\times 0.89443)^2 \div |0.40825|^2 = 1$$

$$x^2\left\{1.11804^2 + 0^2 \div |0.40825|^2\right\} = 1;\ x^2\left\{1.11804^2\right\} = 1;\ 1.11804x = \pm 1$$

By taking the positive, one obtains: $x = 0.89442$; $y = -0.44721$; $z_9 = -\sqrt{1 - x^2 - y^2} = -0.00401 \approx 0$. It coincides with $z = 0$ of $\varphi_0 = 2.67795$. Therefore, the computation is reasonable.

8.3. General Description for the Project Computation on Unit Sphere

The equation of decent slip plane P is z=ax+by, whose normal orientation is: $\boldsymbol{n} = (a, b, -1)$. From (a,b), one obtains $\operatorname{actg}\frac{b}{a}$. Suppose q denotes the steepest ascent angle in the horizontal plane {Oxy}, then by using (a,b) one obtains $\theta = \operatorname{actg}\frac{b}{a} \vee \theta = \pi + \operatorname{actg}\frac{b}{a}$. It follows the steepest decent angle $\theta^T = \theta \pm 180^\circ = \theta \pm \pi$, the allowed interval is $\Delta^T = \left(\theta^T - \frac{\pi}{2}, \theta^T + \frac{\pi}{2}\right)$. The equation of unit sphere is

$$x^2 + y^2 + z^2 = 1 \tag{103}$$

Let $\rho = \sqrt{a^2 + b^2}$, $a = \rho\cos\theta$, and $b = \rho\sin\theta$, then after horizontal project normalized the steepest ascent orientation is denoted by $s_0 = \frac{s}{\rho}$, which may be expressed by $s_0 = (\cos\theta, \sin\theta, \rho)$, $\cos\theta = \frac{a}{\rho}$, and $\sin\theta = \frac{b}{\rho}$. The ascent angle is $\alpha = \operatorname{actg}\rho$

In the plane $z=0$, we consider the following ellipse, whose long radius is equal to 1, and short radius equal to |cosa|. The orientation of long radius is x, whose orientation denotes to the angle $\theta^T+\frac{\pi}{2}$. The orientation of short radius is h, whose orientation points to q. In the coordinates {xOh}, the equation of the ellipse is as follows

$$\xi^2+\frac{\eta^2}{|\cos\alpha|^2}=1 \quad (104)$$

We want to find the equation of the ellipse in the coordinates {xOy}. Refer to the **Figure**ure 6, in which y denotes the angle from Ox axis to Ox axis. The coordinate transformational formula is $\zeta = x\cos\psi - y\sin\psi$; and $\eta = x\sin\psi + y\cos\psi$. Substituting into Equation (104), one obtains

$$(x\cos\psi - y\sin\psi)^2 + \frac{(x\sin\psi + y\cos\psi)^2}{|\cos\alpha|^2}=1 \quad (105)$$

Let $x=\cos\varphi$, $y=\sin\varphi$, then $z=a\cos\varphi+b\sin\varphi$, the allowed interval is $\Delta^T=\left(\theta^T-\frac{\pi}{2},\theta^T+\frac{\pi}{2}\right)$. Now we divide Δ^T into equidistance 10^n sub-intervals; i.e. we take a set $\{\varphi_i\}$, where $i=0,1,\cdots,10^n$. Let $\kappa=10^n$, and let $\delta=\frac{\pi}{\kappa}$, $\varphi_{i+1}=\varphi_i+\delta$, $\varphi_0=\theta^T-\frac{\pi}{2}$, $\varphi_k=\theta^T+\frac{\pi}{2}$.

The computational process: for given φ_i, we firstly compute the orientation vector $(\cos\varphi_i,\sin\varphi_i)$. According to its signs, we define what quadrant does the solution (x_i,y_i) fall in, or what axis it falls on. Then by using the normal equation: $-x\sin\varphi_i+y\cos\varphi_i=0$; we determine $y=x\mathrm{tg}\varphi_i$. Substituting into the ellipse in Equation (105), one obtains

$$(x\cos\psi - y\sin\psi)^2 + \frac{(x\sin\psi + x\mathrm{tg}\varphi_i\cos\psi)^2}{|\cos\alpha|^2}=1 \quad (106)$$

From the above formula and refer to quadrant what the solution (x_i,y_i) falls in, or what axis the solution falls on, then we solve the value of x_i. Then we compute: $y_i=x_i\mathrm{tg}\varphi_i$. And by using the formula

$$z_i=-\sqrt{1-x_i^2-y_i^2} \quad (107)$$

we compute the value z_i. The three dimensional coordinates (x_i, y_i, z_i) is namely the intersectional point of the plane P and the unit sphere along the orientation angle φ_i.

For the set $\{\varphi_i\}$, after finding the corresponding three dimensional points, one obtains a downward semi circular plane. The solid angle between it and the horizontal plane is namely the domain of alloweDecent orientation.

For the discussion of m planes, refer to Section 7, here we don't repeat it.

8.4. Complements for Vertical Plane

In view of theory, it is possible for the existing of a decent slip plane Π_0 which is perpendicular to the horizontal plane. Suppose the equation of Π_0 is $\alpha x + \beta y = 0$, where $(\alpha, \beta) \neq (0,0)$. Its normal has two possibilities: $\boldsymbol{v}_1 = (\alpha, \beta, 0)$, and $\boldsymbol{v}_2 = (-\alpha, -\beta, 0)$, in which there is only one who points to the solid part of Π_0, denoted by: $\boldsymbol{n} = (a, b, 0)$. Then it holds: $(\boldsymbol{n} = \boldsymbol{v}_1) \vee (\boldsymbol{n} = \boldsymbol{v}_2)$. The equation of Π_0 turns into: $ax + by = 0$, i.e. $x\cos\theta + y\sin\theta = 0$.

From (a,b) it follows $\mathrm{actg}\frac{b}{a}$. Suppose q is the steepest ascent angle in horizontal plane {Oxy}, then from the signs of (a,b) we are able to determine

$\theta = \mathrm{actg}\frac{b}{a} \vee \theta = \pi + \mathrm{actg}\frac{b}{a}$. Then we solve the steepest decent angle $\theta^T = \theta \pm 180^\circ = \theta \pm \pi$, and the allowed interval $\Delta^T = \left(\theta^T - \frac{\pi}{2}, \theta^T + \frac{\pi}{2}\right)$

Let y denote the angle from Ox axis to Ox axis, whose value is equal to the angle from $\theta^T + \frac{\pi}{2}$ to Ox axis. We have the following coordinate transformational formula

$$\xi = x\cos\psi - y\sin\psi;\ \eta = x\sin\psi + y\cos\psi \quad (108)$$

In the coordinates {Oxhz}, the equation of unit sphere is as follows

whose project G on the plane $\eta = 0$ is as follows:

$$\xi^2 + z^2 = 1 \quad (110)$$

Transforming into the coordinates {Oxyz}, the equations of G is as follows

$$(x\cos\psi - y\sin\psi)^2 + z^2 = 1 \tag{111}$$

and

$$x\sin\psi + y\cos\psi = 0 \tag{112}$$

When $|\cos\psi| \geq |\sin\psi|$, one solves

$$y = -x\sin\psi \div \cos\psi \tag{113}$$

Substituting into Equation (111) one obtains

$$(x\cos\psi + \sin\psi \times x\sin\psi \div \cos\psi)^2 + z^2 = 1 \tag{114}$$

$$(x \div \cos\psi)^2 + z^2 = 1 \tag{115}$$

This is an ellipse whose long radius is equal to 1 and on z axis, and whose short radius is equal to |cosy| and on x axis. This is a formula expression in coordinate {Oxz} of G which is the project of the great circle of unit sphere on the plane $\eta = 0$. We solve

$$|x| \leq |\cos\psi| \tag{116}$$

When $x \leq \cos\psi$, $y = -\sin\psi$, $z = 0$. It coincides. When $x = 0$, $y = 0$, $z = \pm 1$, this is the steepest decent slip orientation. It coincides.

When $|\cos\psi| \leq |\sin\psi|$, one solves

$$x = -y\cos\psi \div \sin\psi \tag{117}$$

Substituting into Equation (111) one obtains

$$(\cos\psi \times y\cos\psi \div \sin\psi + y\sin\psi)^2 + z^2 = 1 \tag{118}$$

$$(y \div \sin\psi)^2 + z^2 = 1 \tag{119}$$

where is an ellipse with its long radius equal to 1 and on z axis, and its short radius equal to $|\sin\psi|$ and on y axis. This is a formula expression in coordinate {Oyz} of G which is the project of the great circle of unit sphere on the plane h = 0. We solve

$$|y| \le |\sin\psi| \tag{120}$$

When $x=\cos\psi$, $y=-\sin\psi$, $z=0$. This coincides. When $x=0$, $y=0$, $z=\pm1$. This is the steepest decent slip orientation. It also coincides.

From formulas Equation (113), Equation (115) and satisfying the condition Equation (116), for every given x value, we may compute z value for y = 0. Therefore, we can draw the locus of G.

In order to solve the solid angle between F and G, where F is the horizontal semi circle of unit circle $x^2+y^2=1$ along the orientation θ^T, let $x=\cos\psi$, $y=\sin\psi$, $\theta^T-\frac{\pi}{2}\le\varphi\le\theta^T+\frac{\pi}{2}$, then we take a set $\{\varphi_i\}$, where $i=0,1,\cdots,10''$. Let $\kappa=10''$, and let $\delta=\frac{\pi}{\kappa}$, $\varphi_{i+1}=\varphi_i+\delta$, $\varphi_0=\theta^T-\frac{\pi}{2}$, $\varphi_k=\theta^T+\frac{\pi}{2}$. According to the coordinates (x,y), one may draw the **Figure** of F.

Because the data of the **Figure** of vertical plane has solved, it follows that the mathematical method for spherical project of landslip plane is also obtained. For the detail of **Figure** , please refer to three dimension software. The author must do an optimal stopping now.

REFERENCES

1. S. G. Hua, "Stereographic Projection of Rock Stable Analysis," Scientia Sinica, Vol. 3, No. 5, 1977, pp. 260-272.
2. a. Shokichi, "Encyclopedic Dictionary of Mathematics," Iwanami Shoten, Tokyo, 1954, p. 459.
3. W. G. Lei and T. Chou, "A Short Course of Analytical Geometry," 2nd Edition, High Education Press, Beijing, 2007.

CHAPTER 5

Pointe Drawings of Planar Graphs

Oswin Aichholzer[a], Günter Rote[b], André Schulz[c], Birgit Vogtenhuber[a]

[a] *Institute for Software Technology, Graz University of Technology, Austria*

[b] *Institut für Informatik, Freie Universität Berlin, Germany*

[c] *Institut für Mathematsche Logik und Grundlagenforschung, Universität Münster, Germany*

ABSTRACT

We study the problem how to draw a planar graph crossing-free such that every vertex is incident to an angle greater than π. In general a plane straight-line drawing cannot guarantee this property. We present algorithms which construct such drawings with either tangent-continuous biarcs or quadratic Bézier curves (parabolic arcs), even if the positions of the vertices are predefined by a given plane straight-line drawing of the graph. Moreover, the graph can be drawn with circular arcs if the vertices can be placed arbitrarily. The topic is related to non-crossing drawings of multigraphs and vertex labeling.

Keywords: Planar drawings; Pointed pseudotriangulations; Angular resolution

1. INTRODUCTION

Throughout this paper, let $G = (V, E)$ be a simple planar graph without loops, with finite vertex set V and thus a finite set of edges E. We use the natural understanding of a drawing of a graph. Vertices are represented as points in the plane and edges as continuous and (at least piecewise) differentiable curves connecting the points of adjacent vertices. A drawing is called non-crossing or plane, if the drawn edges do not intersect in their interior. If we consider only topological properties, that is, the order of the edges and consequently of the faces, we refer to this as combinatorial embedding. Given a (combinatorial) embedding of a graph G, the faces of G are defined as usual. By an (abstract) plane graph, we mean an equivalence class of plane drawings under homotopic deformations of the plane. For connected graphs, this amounts to a combinatorial embedding together with a designation of the outer face.

For a drawing $\mathcal{F}(G)$ of G we denote the placement of a vertex $v \in V$ by $\mathcal{F}(v)$, by $\mathcal{F}(v)$, and the drawing of an edge $e \in E$ by $\mathcal{F}(e)$. Note that we consider embedded edges to be open, i.e., the do not contain their endpoints. For simplicity, and as there is no risk of confusion, in the **Figure**ures we will denote embedded vertices just by v instead of $\mathcal{F}(v)$.

The tangent of an edge $\mathcal{F}(e)$ at a vertex $\mathcal{F}(v)$. is the limit of the tangents to $\mathcal{F}(e)$ when approaching $\mathcal{F}(v)$. along $\mathcal{F}(e)$ The tangent ray of $\mathcal{F}(e)$ at $\mathcal{F}(v)$. is the open ray along the tangent to $\mathcal{F}(e)$ at $\mathcal{F}(v)$. from F(v)F(v) towards $\mathcal{F}(e)$ A drawing gives us a cyclic order of incident edges around each vertex. The angle between two consecutive edges incident to a vertex $\mathcal{F}(v)$. is defined as the angle between the corresponding tangent rays at $\mathcal{F}(v)$. that does not contain the tangent ray of any other incident edge. We say that this angle is incident to $\mathcal{F}(v)$. (and vice versa). In the case of a degree two vertex there are two such angles between the two incident edges. If a vertex has degree at most one, we say that it is incident to one angle (having value 2π).

Definition 1 Pointedness.

A vertex in a drawing $\mathcal{F}(G)$ is called pointed if it is incident to an angle greater than π (see **Figure**. 1). We say that a vertex is pointed to a face if its large angle lies in this face. If all vertices in a drawing are pointed we call the drawing pointed.

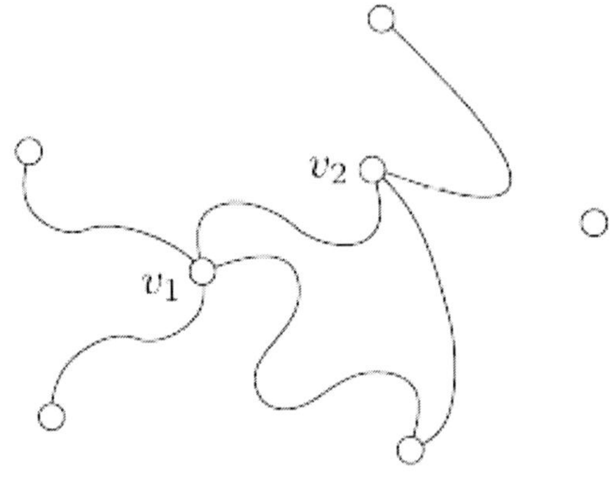

Figure 1. Drawing with a non-pointed vertex v1v1 and a pointed vertex v2v2.

For the special case of straight-line drawings, this definition is identical to the classic definition of pointedness, a term which stems from the field of pseudotriangulations. A pseudotriangle is a simple polygon with exactly three vertices with interior angle smaller than π. A pseudotriangulation is a plane straight-line graph where every interior face is a pseudotriangle and the outer face is convex. Pseudotriangulations have rich applications and are an important geometric data structure, see for example [14], [15] and [18], and the survey [16].

The graphs which can be drawn as pseudotriangulations are well-characterized: A graph is called generically rigid, if its straight-line realization on a generic point set induces a rigid framework (edges represent fixed length rods and vertices represent joints). In two dimensions, there exists an easy combinatorial characterization of generically rigid graphs that become non-rigid after removing an arbitrary edge [10]. These graphs are called Laman graphs. Due to Streinu [18], a graph of a pointed pseudotriangulation is a Laman graph. Conversely, as observed by Haas et al. [8], every plane Laman graph can be realized as a pointed pseudotriangulation. As a consequence, subsets of plane Laman graphs are exactly the graphs that admit a pointed non-crossing straight-line drawing. An operation that preserves the Laman property is the so-called Henneberg operation of type 1: Adding a new vertex of degree 2 to an existing Laman graph will create another Laman graph. A simple example of a planar graph that has no pointed straight-line drawing without crossings is the complete graph with four vertices.

We consider various incarnations of the problem how to draw a plane graph pointed, using different kinds of edge shapes. With arbitrary smooth curves or polygonal chains, the task of constructing a pointeDrawing of a given plane graph is trivial. As natural, but still quite simple edge shapes, we study circular arcs, tangent continuous biarcs, and quadratic Bézier curves. Let us briefly review the definition and basic properties of these curves. A tangent continuous biarc consists of two circular arcs that are joined in a way that they form a C1C1 continuous curve. A quadratic Bézier curve b spanned by three points p1, pmpm and p2 is defined by the equation

$$b(t) = (1-t)^2 p_1 + 2t(1-t)p_m + t^2 p_2, \quad t \in [0, 1].$$

It lies completely inside the triangle p1pmp2p1pmp2 (which is also called control polygon of b), has p1 and p2 as endpoints, and is tangent to p1pm at p1 and to p2pmp2pm at p2. The class of quadratic Bézier curves is the same as the class of parabolic arcs.

We also consider the “extent of pointedness”. For example, can we guarantee a free angular space around each vertex bigger than a given fixed angle larger than π? For this stronger pointedness criterion we define the term ε-pointedness.

Definition 2 ε-Pointedness.

Let $\varepsilon>0$ be a real number. A vertex in the drawing $\mathcal{F}(G)$ is called ε-pointed if it is incident to an angle greater than $2\pi-\varepsilon$ We call a drawing ε-pointed if every vertex is ε-pointed.

In other words, all edges incident to an ε-pointed vertex emanate in a sector of angle ε.

Further, we propose a stronger version of the pointeDrawing problem: Given a plane straight-line drawing $\mathcal{F}_S(G)$, can we redraw it as a plane pointeDrawing with a certain family of edge shapes without moving the vertices? We call a drawing with this property a pointed redrawing. The motivation of a pointed redrawing is clear: we can benefit from the given drawing and keep its advantages (e.g., all vertices are placed on an integer grid or fulfill other optimality criteria).

A more general redrawing problem would start from a plane embedding with not necessarily straight edges. We have not considered this question.

1.1. Results

In Section 2, we consider the problem of pointed redrawings . We show that every plane straight-line drawing $\mathcal{F}_S(G)$,can be redrawn pointed and plane with Bézier curves as well as with tangent continuous biarcs. We show that this is not always possible with circular arcs as edges.

Section 3 then deals with pointeDrawings of (abstract) plane graphs. We prove that every plane graph can be drawn ε -pointed with Bézier curves, for arbitrary small $\varepsilon>0$ We show that by using biarcs as edges, every plane graph can be drawn such that for all vertices v, all incident edges share a common tangent ray at v. This is maybe one of the most beautiful results in this paper from an aesthetic point of view. We further prove that every plane graph can be drawn pointed and plane with circular arcs as edges. For pointeDrawings with biarcs, Bézier curves, or polygonal chains of length two, we give a tight bound on the number of edges that have to be drawn as non-straight curves (Theorem 3.6).

We summarize the results presented in this paper in Table 1. All obtaineDrawings can be constructed algorithmically, with the exception of the methoDescribed in the proof of Theorem 3.2, which needs a disk packing of the plane graph in a preprocessing step.

Table 1. Results presented in this paper.

Edge shape	Problem instance	Obtained result
Circular arcs	Pointed drawing	Possible, Theorem 3.5
	Pointed redrawing	Not possible, Theorem 2.3
Tangent continuous biarcs	ε-pointed drawing	Possible, Theorem 3.2
	Pointed redrawing	Possible, Theorem 2.2
Quadratic Bézier curves	ε-pointed drawing	Possible, Theorem 3.1
	Pointed redrawing	Possible, Theorem 2.1

1.2. Related work and applications

Traditionally, graph drawing is mainly concerned with using the simplest class of curves for the edges: straight-line segments. According to Fáry's theorem [7], every (simple) plane graph has a plane straight-line drawing in the Euclidean plane. There is a vast literature dealing with the question of efficiently finding plane straight-line drawings that fulfill certain (optimality) criteria (see [3] and [12] for an overview). Improving work of de Fraysseix, Pach and Pollack [6], Schnyder [17] proved that every plane graph with n vertices has a plane straight-line drawing where the vertices lie on a grid of size $(n-2) \times (n-2)$. The famous Koebe–Andreev–Thurston circle packing theorem [2] and [9] states that every plane graph can be embedded with straight-line edges in a way such that its vertices correspond to interior disjoint disks, which touch if and only if the corresponding vertices are connected with an edge, see also [13] and [4]. We will use both the procedure of de Fraysseix, Pach and Pollack [6] and circle packings as building blocks in our algorithm.

If we relax the condition that the given plane graph has to be simple, Fáry's theorem does not hold, for the simple reason that straight-line drawings are not well defined for loops, and multiple edges between two vertices are excluded. However, with more complex edge shapes, one can ask for crossing-free drawing of planar multigraphs with loops. The most natural approach is to allow circular arcs. Drawing multiple edges as circular arcs is no problem, as an edge in a straight-line drawing can be perturbed to any number of close-by circular arcs. Loops, however, require more space. The only circular arc between a vertex and itself is a full circle through this vertex. Thus, the vertex has to be incident to an angle of at least π, which then is sufficient for any number of nested loops at this vertex. This means that the simple graph containing only the non-loop edges must be drawn as a pointed graph. In Section 3.2, we show that such a drawing exists (Theorem 3.5) and as a consequence, we obtain a plane drawing with circular arcs for every planar multigraph (Corollary 3.1). Moreover, an $O(n) \times O(n^2)$ grid is sufficient to embed the vertices of these drawings.

Another potential application for constructing pointeDrawings of graphs comes from drawing vertex labels. If the edges incident to a vertex point in all directions, it might be hard to place a label close to its vertex. Thus it is good to have some angular space without incident edges.

These results were presented at the 2007 Canadian Conference on Computational Geometry in Ottawa [1]. One of the results which we announced in the proceedings version cannot be maintained in full generality, see Theorem 3.6 below.

2. POINTED REDRAWINGS

We start with the redrawing problem. Throughout this section we consider a plane straight-line drawing as input of our problem instance. Let this drawing be $\mathcal{F}_S(G)$.

Theorem 2.1. For every plane straight-line drawing $\mathcal{F}_S(G)$ of a simple planar graph G there exists a pointed plane redrawing $\mathcal{F}_q(G)$ with quadratic Bézier curves as edges : $\mathcal{F}_q(v) = \mathcal{F}_S(v)$ for all $v \in V$, and for every vertex $v \in V$, , the cyclic order of the edges incident to v in Fs(G)Fs(G)is the same as in $\mathcal{F}_q(G)$.

Proof. Without loss of generality assume that in $\mathcal{F}_S(G)$ no two vertices have identical x-coordinates or y-coordinates. Assume further, that the vertices are sorted by y-coordinates in increasing order.

We construct $\mathcal{F}_Q(G)$ by iteratively replacing the straight-line edges of $\mathcal{F}_S(G)$ with quadratic Bézier curves. We first replace the edges incident to the bottom-most vertex v1v1, then the edges incident to v2v2, and so on. During the construction we maintain the following two invariants:

(1) For every vertex vivi, the tangent rays of all already redrawn edges lie in the open halfplane H_i^- below the horizontal line through $\mathcal{F}_q(v_i)$. .

(2) The intermediate drawing is plane.

When replacing the edges incident to a vertex vivi, all edges incident to a vertex *vj*, with $j < i$, have already been redrawn by our algorithm (as all vertices below vivi have already been processed). Let $E_i = e_1, \ldots, e_k$ be the edges which have not been replaced yet, sorted by absolute slope, such that e1e1 has the smallest absolute slope. We redraw these edges in increasing order.

Let $e = v_i v_j,\ j > i$ be the current edge we want to process (see **Figure.** 2). Due to invariant (1) and the processing order of the edges incident to vivi we can choose a point pmpm in H_i^- such that the triangle $t=\mathcal{F}_q(v_i)p_m\mathcal{F}_q(v_j)$ does not contain any vertex or part of an edge of the current drawing in its interior. By convention we place p_m to the right of $\mathcal{F}_q(v_i)$ if e has positive slope, otherwise to the left. We use the triangle t as a control polygon for a quadratic Bézier curve b with endpoints $\mathcal{F}_q(v_i)$ and $\mathcal{F}_q(v_j)$, which we take as replacement for the straight-line edge $\mathcal{F}_s(e)$. Note that b is tangent to $\mathcal{F}_q(v_i)p_m$ at $\mathcal{F}_q(v_i)$, and thus invariant (1) still holds for v_i. As b lies completely inside t , and t, and $t\setminus\{\mathcal{F}_q(v_j)\}$ lies completely inside H_j^-, invariant (1) for vjvj and invariant (2) remain fulfilled as well.

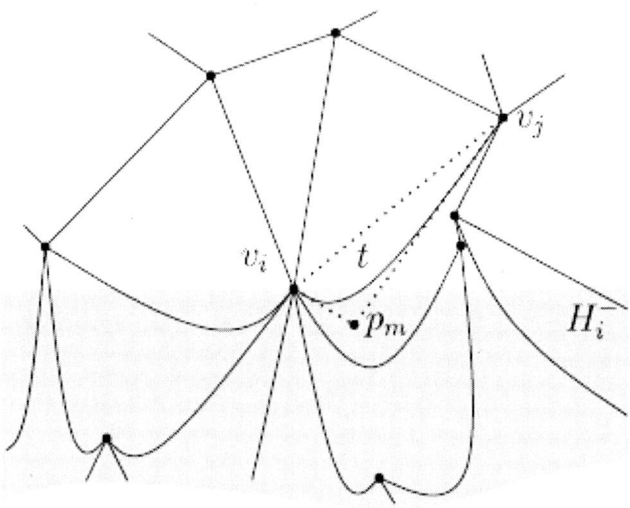

Figure 2. Constructing a plane pointeDrawing where the edges are quadratic Bézier curves (intermediate step).

Having redrawn all edges in this way, we obtain a drawing whose pointedness follows directly from invariant (1), and that is plane follows from invariant (2).

The easiest way to establish that the cyclic order of edges is unchanged is to augment the drawing $\mathcal{F}_s(G)$ to a triangulation by adding edges anDeleting the corresponding arcs in the final drawing. For a triangulation a with fixed outer face the order of the edges around a vertex is unique up to a global reflection [19]. Hence, this order has to be preserved in $\mathcal{F}_q(G)$. □

The technique used in the proof of Theorem 2.1 can be modified to show a similar statement for (tangent continuous) biarcs due to the following observation.

Lemma 2.1. Consider a triangle spanned by three points p1, pmpmand p_2. There exists a tangent continuous biarc connecting p1with $p_2$that lies inside the triangle. Furthermore, the biarc is tangent to p1pmat one end and tangent to p_2p_m at the other end.

Proof. Assume that the segment p1p_m is shorter than p2pmp2pm. The first arc starts in p1with tangent direction p1pm and touches the line p_2p_m in some point This point on the segment p_2p_m has the property that the length $p_m\bar{p}$ of is equal to the length of p1pm (see **Figure**. 3). The center of the arc is found by intersecting the line l1l1 perpendicular to p1p_m through p1 with the line l2l2 perpendicular to p2pmp2pm through $\bar{p}$ The second part of the biarc is given by the straight line segment $\bar{p}p_2$ (a degenerate circular arc). □

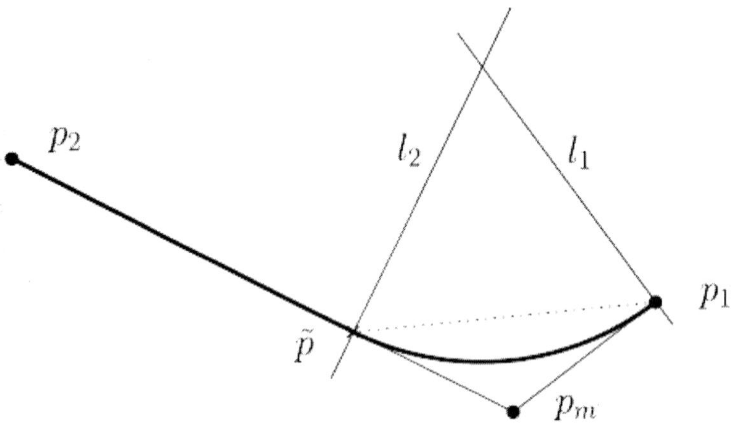

Figure 3. Drawing an edge as a tangent continuous biarc in a triangle.

Theorem 2.2. For every plane straight-line drawing $\mathcal{F}_s(G)$ of a plane graph G there exists a pointed plane redrawing $\mathcal{F}_b(G)$ with tangent continuous biarcs as edges : $\mathcal{F}_b(v) = \mathcal{F}_s(v)$ *for all* $v \in V$, and for every vertex $v \in V$, the cyclic order of the edges incident to v in $\mathcal{F}_s(G)$ is the same as in $\mathcal{F}_b(G)$.

Proof. We re-use the construction from the proof of Theorem 2.1. Whenever we have chosen an appropriate empty triangle for an edge replacement, we place a tangent continuous biarc in it (as described in Lemma 2.1). □

We conclude this section with a negative result on pointed redrawings.

Theorem 2.3. There is a planar graph $G=(V,E)$ with a plane straight-line drawing $\mathcal{F}_s(G)$, for which there are no pointed plane drawings $\mathcal{F}_c(G)$ with circular arcs as edges such that $\mathcal{F}_c(v)=\mathcal{F}_s(v)$ *for all* $v \in V$.

Proof. Consider the graph G shown in **Figure**. 4(a). Vertex vcvc is placed at the origin, vertex v_t at $(0,2)$, vertex vlvl at (−0.2,1)(−0.2,1), and vertex vrvr at (0.2,1)(0.2,1). The positions of the remaining vertices are obtained by rotating these vertices by ±120 degrees. Since G is 3-connected and planar, its combinatorial embedding is fixed for any non-crossing drawing [19]. This implies that in any such drawing the edge between vcvc and vtvt has to pass through the narrow passage between vlvl and vrvr. Since we are restricted to circular arcs, the arc connecting vtvt and vcvc has to lie in the shaded region depicted in **Figure**. 4(b). This region is the intersection of the disk touching vtvt, vlvl, vcvc with the disk touching vtvt, vrvr, vcvc. The region lies inside a wedge of angle α=45.3α=45.3 degrees. Thus, the tangents of two arcs from vcvc to the convex hull are separated by an angle of at most β=165.3 degreesβ=165.3 degrees. But in order to make the vertex vcvc pointed, one of these angles would have to be larger than π. □

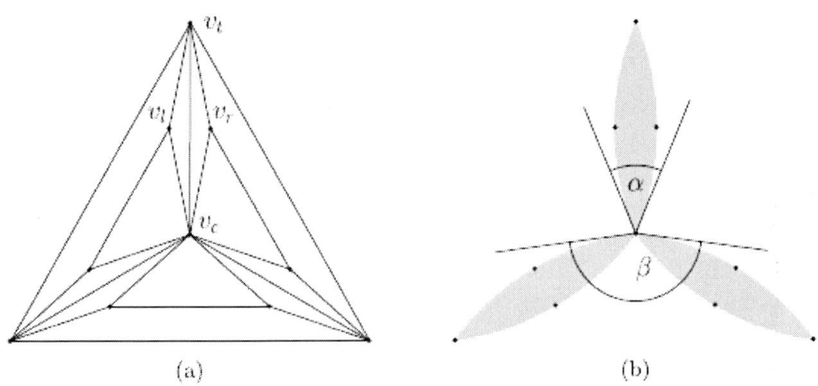

Figure. 4. Example of a straight-line drawing that cannot be redrawn pointed with circular arcs.

Larger examples can be constructed easily. As long as a straight-line drawing similar to **Figure**. 4(a) is contained inside another drawing, a pointed redrawing with circular arcs is impossible. Moreover, with a construction similar to the one shown in **Figure**. 4(a), but with many "spokes" (instead of just three), one can force the largest possible angle free of incident edges at the central vertex to be arbitrary small.

3. POINTEDRAWINGS

3.1. PointeDrawings with Bézier curves and biarcs

In the last section the placement of the points was determined by a given plane straight-line drawing. If the location of the vertices can be chosen arbitrarily, we get the following easy consequence of Theorem 2.1.

Theorem 3.1. For any ε>0 and any plane graph G, there exists a plane drawing Fq(G)Fq(G)with quadratic Bézier curves where all vertices are ε-pointed.

Proof. Consider an arbitrary straight-line drawing Fs(G) In the proof of Theorem 2.1 we showed a construction for a pointeDrawing $\mathcal{F}'_q(G)$ in which for every vertex v and for every edge e incident to v , the tangent ray of $\mathcal{F}'_q(e)$ at $\mathcal{F}'_q(v)$ lies below the horizontal line through $\mathcal{F}'_q(v)$ By compressing the x-axis (i.e., scaling by a factor less than 1), the large angle at every vertex in the resulting drawing increases towards 2π . This modification produces no crossings. Moreover, every quadratic Bézier curve is transformed to a quadratic Bézier curve (with respect to the compressed control polygon). Thus, sufficiently compressing $\mathcal{F}'_q(G)$ results in the desired ε -pointeDrawing Fq(G) □

By similar arguments, it is possible to obtain an ε -pointeDrawing Fb(G) with biarcs. In this case the argumentation is more involved, because compressing a biarc in one direction does not result in another biarc. However, we can modify the proof of Theorem 2.1 in the following way: Recall that we used as invariant (1) that for every vertex vivi, the tangent rays of all already redrawn edges lie in the open halfplane H_i^- To obtain a stronger result, we consider vertical double-wedges centered at the embedded vertices with wedge angle ε , and redefine the region H_i^- to be the wedge below the horizontal line through the embedded vertex. We compress the x-axis until all edges of the compressed straight-line drawing lie strictly within the double-wedges of their endpoints, and apply the previous approach with the changed invariant to this compresseDrawing.

A disadvantage of this construction is that the biarcs tend to consist of a circular arc with small radius and a circular arc with infinite radius. Thus, these drawings are not aesthetically pleasing. For this reason, we present a completely different approach, which also fulfills an even stronger criterion of pointedness. This criterion, namely that all arcs incident to a vertex share a common tangent at this vertex, implies ε -pointedness for any ε>0ε>0.

Theorem 3.2. Every plane graph G=(V,E) has a plane pointeDrawing Fb(G) with tangent-continuous biarcs as edges such that Fb(G)Fb(G)is pointed. Moreover, for every vertex v, all edges incident to v share a common tangent at Fb(v). The directions of these tangents can be independently specified for each vertex.

We emphasize that the locations of the vertices cannot be specified in this theorem.

Proof. According to the Koebe–Andreev–Thurston circle packing theorem [2] and [9], every plane graph admits a disk packing, where each disk belongs to a vertex (which is the center of the disk), and two disks touch if and only if the corresponding vertices share an edge.

We start with such a disk packing of the graph G (see [4], [11] and [5] for algorithmic aspects of such packings). To get our drawing $\mathcal{F}_b(V)$ of the vertices, we place every vertex vivi arbitrarily on the boundary of its disk D_i, avoiding touching points of the disks. The edges incident to vivi will emanate from $\mathcal{F}_b(v_i)$ perpendicular to D_i into the interior of D_i. Thus, we can obtain desired tangent direction for the edges by placing v_i on D_i appropriately. We can avoid the coincidence of vivi with a touching point by rotating the whole disk packing. (There are only finitely many rotations that have to be avoided.)

Now consider an edge $v_i v_j \in E.$ For the embedded vertex Fb(vi)Fb(vi) let titi be the tangent through $\mathcal{F}_b(v_i)$ to its disk D_i. Furthermore, let pijpij be the touching point of the two adjacent disks D_i anD_j and let tijtij be the tangent to DiDi anD_j through pijpij (see **Figure**. 5). We draw a circular arc C_i from $\mathcal{F}_b(v_i)$ to pijpij inside D_i, the center of CiCi being the crossing of titi and tijtij. Similarly, we draw an arc C_j from $\mathcal{F}_b(v_i)$ to pijpij inside D_j, with center $t_j \cap t_{ij}$. Both arcs meet in pijpij with the same tangent (orthogonal to t_{ij}). Therefore, the concatenation of CiCi and CjCj gives a tangent-continuous biarc. We use C_iC_j as drawing for v_iv_j and apply this construction for all edges in E.

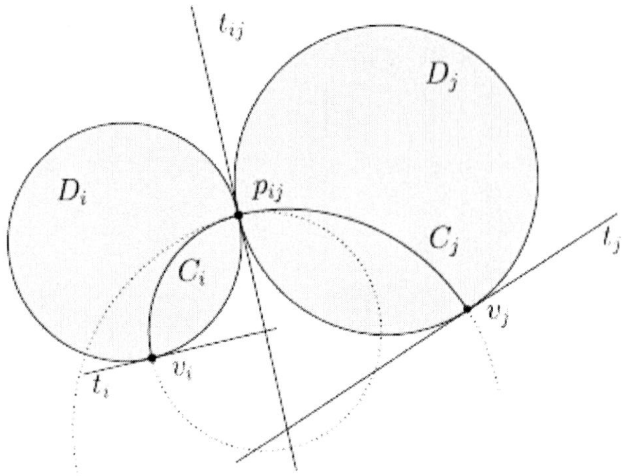

Figure 5. Construction of a tangent-continuous biarc from two touching disks DiDi,

It is left to show that the constructeDrawing is non-crossing. Two biarcs could cross only within a disk of the disk packing. Consider all circular arcs incident to the embedded vertex $\mathcal{F}_b(v_i)$ as depicted in **Figure**. 6. All corresponding circles have their centers on titi and are passing through $\mathcal{F}_b(v_i)$which lies on titi as well. Thus, any two of these circles intersect only in $\mathcal{F}_b(v_i)$and the constructeDrawing is plane.

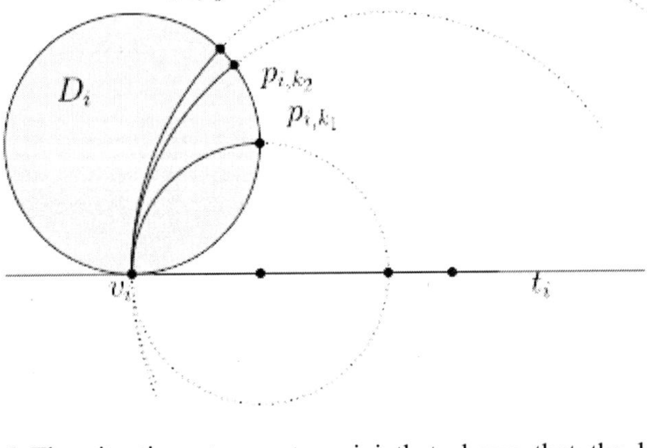

Figure 6. The situation at a vertex vivi that shows that the biarcs do not intersect.

All biarcs incident to an embedded vertex $\mathcal{F}_b(v_i)$ have a common tangent orthogonal to titi. We can determine this tangent by placing the vertex vivi on CiCi appropriately, avoiding the finitely many touching points of DiDi. □

The above proof leaves some freedom to place the vertices on the boundaries of the corresponding disks. If in the drawing $\mathcal{F}_s(G)$ no two disk centers have the same x-coordinate, we can place each vertex on the bottommost point of the boundary of its disk. By this, all biarcs have positive curvature and we have no "S-shaped" biarcs (see **Figure**. 7).

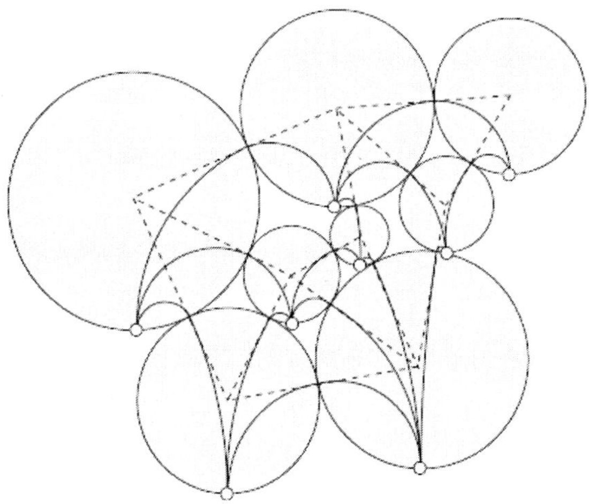

Figure 7. A pointeDrawing with biarcs as edges, constructed from a disk packing.

Another possibility is to place each vertex $v_i \in V$ farthest away from any touching point of its disk DiDi. In this way we can guarantee the radius of any circular arc inside D_i to be at least $R_i \cdot \tan \frac{\pi}{2k_i}$, is the degree of vivi. Unfortunately, in general, we have no control over the radii RiRi in the disk packing.

3.2. PointeDrawings with circular arcs

We assume in this section that no two vertices will get the same y-coordinate in the drawing. The drawing we describe next uses a special type of circular arcs.

Definition 3 Upper horizontally tangent arc, UHT-arc.

Let p1 and p2 be two points, where p1 has the larger y -coordinate. We call a circular arc between p1 and p2upper horizontally tangent if it has a horizontal tangent at p1.

Definition 4 Upper horizontally tangent triangle, UHT-triangle.

We call a drawing of a triangle upper horizontally tangent if all of its edges are drawn as UHT-arcs (see **Figure**. 9).

For any two points, the UHT-arc is uniquely defined. Hence, for every point triple the UHT-triangle is unique. The following lemmata show that under certain assumptions the UHT-triangles behave nicely.

Lemma 3.1. Consider the UHT-arc μ between p1and p2. Let $h_l$be the horizontal line through p1. Then the angle at p1between $h_l$and μ is twice as large as the angle at p1between $p_1p_2$and h_l.

Proof. The situation stated in the lemma is depicted in **Figure**. 8. Let α be the angle at p_1 between h_l and p_1p_2. This angle is the alternate angle to the angle at p_2 between h_2 and p_2p_1. Let ptpt be the intersection of the tangents of μ at p_1 and p_2. The triangle p1p2ptp1p2pt is isosceles and hence the angle between p_1p_2 and p_1p_t is α as well. Thus, the angle between μ and p_1p_2 equals 2α. □

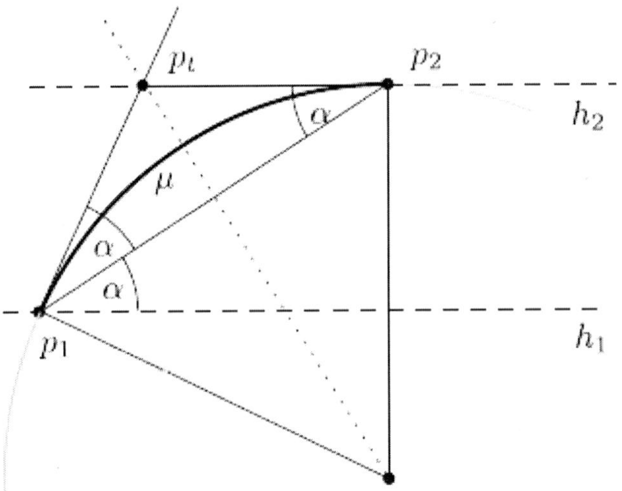

Figure 8. Construction used in the proof of Lemma 3.1.

In the following lemma, we restrict the straight-line edges to have an absolute slope less or equal 1. This implies that the angle between the tangent of an UHT-arc at the lower point and the horizontal line through this lower point is at most π/2π/2. As a consequence, the UHT-arc is x-monotone and is contained in the axis-parallel bounding rectangle spanned by its endpoints.

Lemma 3.2.

Consider three points p_1,p_2,p_3 sorted by their x-coordinates. If

(i)the absolute slope of the line segments p_1p_2, $p_2p_3$and $p_1p_3$is smaller than 1, and

(ii)$p_2$lies below the line through $p_1$and p_3, or p2has the highest y-coordinate,

then p_1, p_2, and $p_3$span a non-crossing UHT-triangle that is oriented in the same way as the straight-line triangle $p_1p_2p_3$. That is, the clockwise order of the points around the faces is the same.

Proof. Let yiyi be the y -coordinate of pipi, let hihi denote the horizontal line passing through pipi, and let aijaij denote the UHT-arc between pipi and p_j.

We prove the lemma by case distinction. Without loss of generality we assume that y1<y3y1<y3. Depending on the relative location of y2y2 we obtain three cases (see **Figure**. 9).

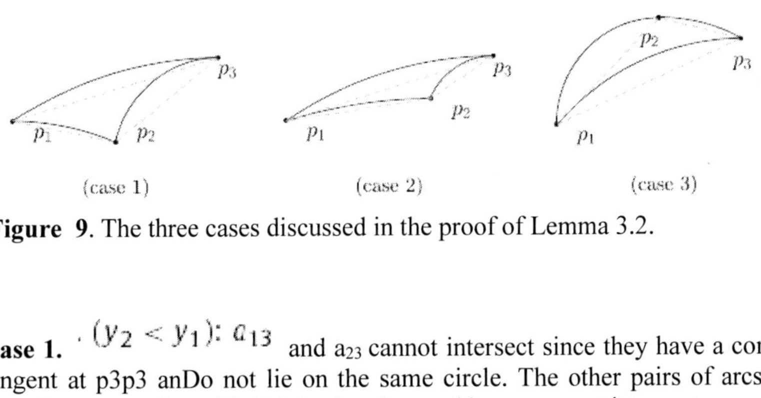

Figure 9. The three cases discussed in the proof of Lemma 3.2.

Case 1. $(y_2 < y_1)$: a_{13} and a_{23} cannot intersect since they have a common tangent at p3p3 anDo not lie on the same circle. The other pairs of arcs have bounding rectangles with disjoint interior, and hence cannot intersect.

Case 2. $(y_1 < y_2 < y_3)$: Again, a_{13} and a_{23} do not intersect since they have a common tangent at p_3 anDo not lie on the same circle. The arcs a_{12} and a_{23} have bounding rectangles with disjoint interior, and therefore do not intersect either. Since p2 lies below the line segment p_1p_3 (condition (ii)), p2 lies below the arc a_{13} and p_1p_3 has larger slope than p_1p_2. Thus, anDue to Lemma 3.1, the angle between the tangent of a_{13} and h_1 is larger than the angle between the tangent of a_{12} and h_1, meaning that a12a12 is incident to p1 "below" a_{13}. As the second endpoint of p_2 of a_{12} lies below a_{13} as well, an intersection of a12a12 and a13a13 (to the right of p1) would imply a second such intersection. This is impossible, because the two circles induced by a12a12 and a13a13 would intersect three times.

Case 3. $(y_3 < y_2)$: The pairs $a23/a12$, and a_{23}/a_{13} have bounding rectangles with disjoint interior and therefore do not intersect. For the remaining pair of arcs $a12$ and a_{13} we apply again Lemma 3.1 and observe that a_{12} is incident to $p1$ "above" a_{13} . As the second endpoint of p_2 of a_{12} lies above $a13$ as well, it follows that an intersection of a_{12} and a_{13} (to the right of p_1) would again imply that the two circles induced by a_{12} and a_{13} intersect three times, which is impossible.

Since in all three cases, the above-below order of the (x-monotone) incident edges at each vertex is preserved, the orientation of the UHT-triangle is identical to the orientation of the straight-line triangle. ✷

We continue by constructing a straight-line drawing that allows us to substitute its triangles by UHT-triangles. The basic idea goes back to de Fraysseix, Pach and Pollack [6].

Theorem 3.3. (See [6].) A plane triangulated graph has a plane straight-line drawing on a $(2n-4)\times(n-2)$ grid.

Let us briefly review the incremental construction used in [6], see **Figure**. 10. The vertices are inserted in a special (so-called canonical) order, such that the next vertex $pk+1$ that is inserted can be drawn on the outer face of the graph Gk induced by the first k vertices. Thereby as invariant it is maintained that the outer boundary of the graph Gk (drawn so far) forms a chain of pieces of slope ±1, resting on a horizontal basis (**Figure**. 10(a)). The next vertex $pk+1$ to be drawn is adjacent to a continuous subsequence of vertices on the outer boundary. To make space for the new edges incident to $pk+1$, the boundary of Gk is split into three pieces, which are separated from each other by shifting them one unit apart (**Figure**. 10(b)). The middle piece contains all neighbors of $pk+1$ except the first and the last one. In [6] it is shown that one can split the upper boundary of Gk at an arbitrary point and shift the pieces apart horizontally, by an arbitrary amount. If an appropriate part of Gk inside the shaded area is shifted along, no crossings are created. Any number of these shifting operation can be carried out in succession. Furthermore, during such a shifting operation, the endpoints of an edge can only be moved farther apart horizontally.

We slightly modify this inductive procedure to prove the following theorem

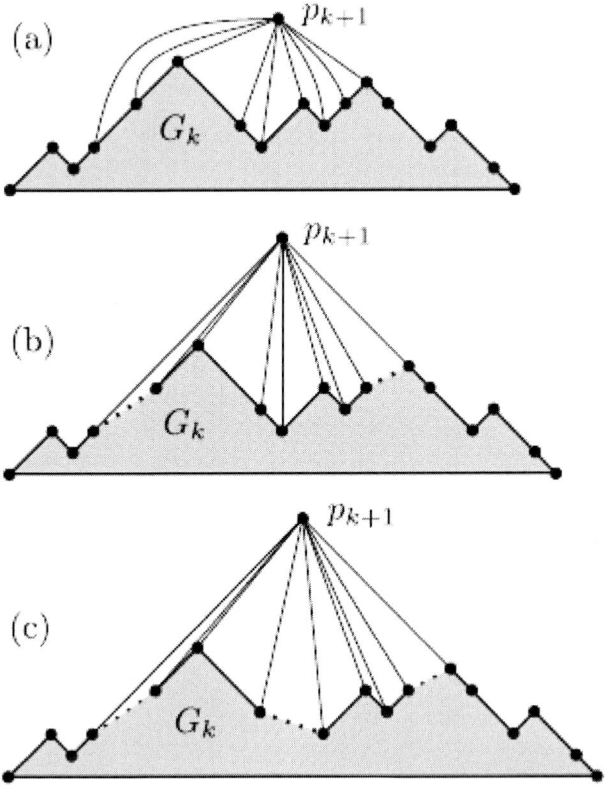

Figure 10. (a)–(b) The incremental step in the straight-line drawing algorithm of de Fraysseix, Pach and Pollack [6], and (c) the modification that prevents vertical edges.

We slightly modify this inductive procedure to prove the following theorem.

Theorem 3.4. A plane triangulated graph has a plane straight-line drawing on a grid of size $(4n-9)\times(2n-4)$, with the following additional properties:

(a)No edge is vertical.

(b)No edge is horizontal.

(c)In each triangular face, the vertex with the middle x-coordinate is either the vertex with the highest y-coordinate, or it lies below the opposite edge.

Proof. The newly created triangles in the incremental construction described above always fulfill property (c), which can be checkeDirectly, and no horizontal edges are created (property (b)). The only horizontal edge is the bottom base edge. This horizontal edge can easily be avoided by starting the construction with a non-horizontal base triangle in the first step.

To prevent vertical edges, one can split the middle part into two pieces by vertical line through pk+1pk+1 and set them apart by two more units (**Figure.** 10(c)). A boundary vertex on the vertical line can be assigned to either part. (Two units are necessary to ensure that the left and right part are separated in total by an even offset; this guarantees that the position of pk+1pk+1, which is defined by the requirement that its leftmost and rightmost incident edges have slope +1 and −1 respectively, gets integer coordinates.)

Adding a vertex preserves the old y-coordinates and the order of the x - coordinates between adjacent vertices, as well as the cyclic order of the edges. As a consequence, properties (b) and (c) can be guaranteed to hold for previously added vertices after shifting. Property (a) is preserved because shifting decreases the absolute slope of an already inserted edge, and by the same reasoning, no edge becomes vertical. The dimensions of the grid increase by 4×24×2 units for each new vertex. The initial drawing of the graph G3G3 with the first three vertices needs a 3×23×2 grid. □

We continue with the main result of this section.

Theorem 3.5. Every plane graph G has a plane pointeDrawing with circular arcs as edges.

Proof. We assume that the graph G is a triangulation. (Otherwise we add edges such that G becomes a triangulation anDelete these edges in the end.) Given an n-vertex plane triangulated graph, the algorithm of Theorem 3.4 constructs drawings in which for every edge its absolute slope is less than 2n. By scaling the x-axis by a factor of 2n, we obtain a drawing in which all edges have slopes in the range between −1 and +1. In this scaleDrawing, all triangles fulfill the conditions of Lemma 3.2. We substitute every straight-line edge by its corresponding UHT-arc. By this substitution, the order of the edges around a vertex is preserved, and every straight-line triangle is replaced by its corresponding UHT-triangle. Thus, anDue to Lemma 3.2, the obtained circular drawing is crossing-free (edges on the upper hull are non-crossing as they have bounding-rectangles with disjoint interior).

Around every vertex there is a number of edges that emanate in the horizontal direction, plus a number of additional edges that point upward. The latter type of edges have distinct tangent directions. Thus one can slightly bend every edge upward and achieve a pointeDrawing with circular arcs. □

Due to Theorem 3.4, pointeDrawings constructed as above lie on an $O(n) \times O(n^2)$ grid. An example of such a drawing is shown in **Figure**. 11.

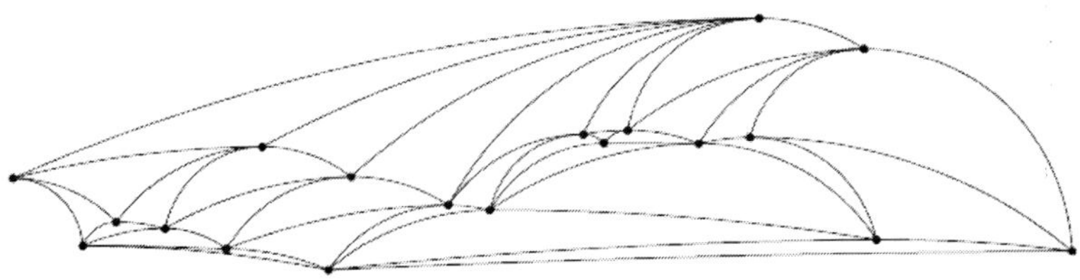

Figure 11. An example of a pointeDrawing with circular arcs. The horizontal stretch factor was chosen just sufficiently to ensure that all straight edges have absolute slope less than 1, instead of 2n.

As a consequence, we obtain the following result about multigraphs with loops, as mentioned in the Introduction:

Corollary 3.1. Every planar multigraph, possibly with loops, admits a plane drawing with circular arcs, whose vertices lie on an $O(n) \times O(n^2)$ grid.

Note that this is no longer true if we insist on a particular combinatorial embedding. For example, we cannot have three non-nested loops incident to a vertex.

3.3. PointeDrawings with the help of combinatorial pseudotriangulations

A different way to find a pointeDrawing uses the framework established by Haas et al. [8]. Let us recall some terminology first. A combinatorial pseudotriangulation is a planar combinatorial embedding of an (abstract) connected planar graph G with an assignment of the tags big/small to the angles of G such that the following conditions are fulfilled:

(1)Every interior face has exactly three small angles.

(2)The outer face has only angles labeled big.

(3)Every vertex is incident to at most one angle labeled big. If it is incident to a big angle, it is called pointed (in the face where is has its big angle).

(4)A vertex of degree at most 2 is incident to one angle labeled big.

An angle assignment that fulfills these conditions is called cpt-assignment.

By [8, Theorem 6], every combinatorial pseudotriangulation whose underlying graph is a Laman graph can be embedded as a pseudotriangulation such that every angle with tag big is larger than π in the drawing, and every angle with tag small is smaller than π in the drawing. Furthermore, the shape of every face can be specified up to affine transformations [8, Theorem 11].

Lemma 3.3. Every triangulation with n vertices can be turned into a combinatorial pointed pseudotriangulation by subdividing n−3n−3edges, each with exactly one additional vertex. Furthermore, the underlying graph is a Laman graph.

Proof. We construct a cpt-assignment by an iterative procedure that is guided by the canonical order of the plane graph (see Theorem 3.3). The assignment ensures that all graphs G_k have a valid cpt-assignment. This can be easily made true for G_3, which is a single triangle. Here, the three angles at the boundary get the tag *big*, and the three interior angles get the tag *small*. Assume now that we add the vertex v_{k+1} to G_k to obtain G_{k+1}. We therefore connect v_{k+1} with its neighbors on the boundary of G_k (see **Figure**. 10(a)). All new edges that are not on the boundary of G_{k+1} are subdivided by adding a new vertex on each such edge. The cpt-assignment is extended such that every exterior angle of G_{k+1} gets the tag *big*, and all newly added faces have exactly three small angles. Every vertex that vanishes from the boundary will now realize its big angle in one of the newly created faces. The simple scheme how to assign the angles is depicted in **Figure**. 12. From a different perspective we add the vertex v_{k+1} by linking it to G_k by two new edges. This operation is a Henneberg-1 step and it preserves the Laman property of the graph. The vertices that are introduced by subdividing edges are a result of additional Henneberg-1 steps. Since G_3 is a Laman graph, also $G_n = G$ is a Laman graph.

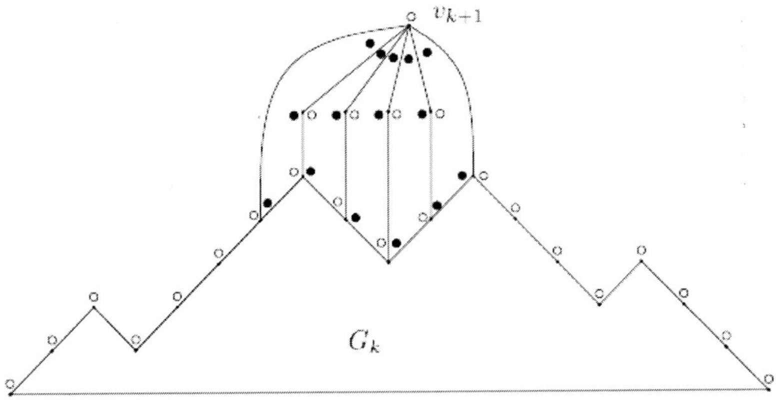

Figure 12. Extending the cpt-assignment during the construction of G by its canonical order.

In every step we add two edges that are not subdivided, and there are n−4 steps necessary to go from G_3 to G_n. Thus, in total, we add 2n−3 edges that are not subdivided. Since the triangulation G has 3n−6 edges, n−3of them are subdivided.

Theorem 3.6. Every plane graph G with n vertices has a plane pointeDrawing with either quadratic Bézier curves, tangent continuous biarcs, or 2-chains (polygonal chains consisting of two line segments), which uses at most n−3n−3non-straight edges.

Proof. We assume that the graph G is a triangulation. (Otherwise we add edges such that G becomes a triangulation anDelete these edges in the end.) As a first step, we turn G into a combinatorial pointed pseudotriangulation as done in Lemma 3.3. By this we creates four types of bounded faces:

(i)triangles,

(ii)quadrilaterals with a degree-2 vertex with big angle,

(iii)quadrilaterals with a degree-2 vertex with small angle and the big angle is realized next to it,

(iv)pentagons with two non-adjacent degree-2 vertices, one of them with a big, one of them with a small angle.

We apply the algorithm of [8] to realize the combinatorial pseudotriangulation. In all faces of type (iii) the interior of the triangle spanned by the degree-2 vertex and its two neighbors is empty, see **Figure**. 13(b). The same is not necessarily true for the faces of type (iv). However, the algorithm of [8] allows us to specify the face shapes up to affine transformations. By giving all faces of type (iv) the shape shown in **Figure**. 13(a) one assures that the interior of shaded triangle in **Figure**. 13(c) is empty. This property is preserved under affine transformations.

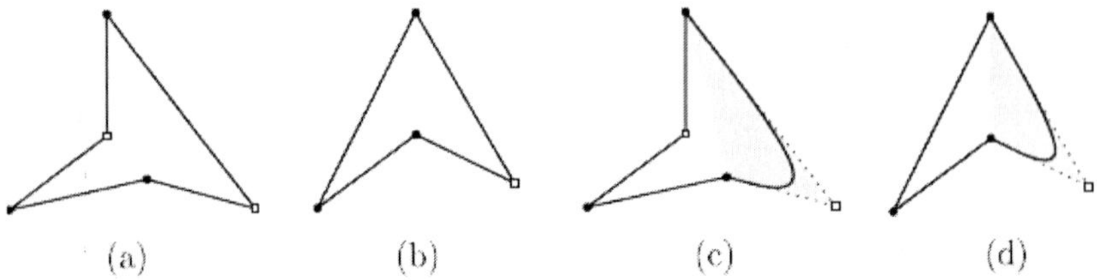

Figure 13. Affine shapes of faces used for the drawing (a)–(b) and control triangles for curve replacement inside these faces (c)–(d). The degree-2 vertices that came from edge subdivisions are marked as boxes.

What we have obtained so far is a pointeDrawing, where at most n−3n−3 edges are drawn as 2-chains, which proves the theorem for the case of polygonal chains.

For the case of Bézier curves or biarcs, we consider for each 2-chain p1, pmpm, p2 (with pmpm being the vertex of degree two) the triangle Δ=p1pmp2 Δ lies in a pseudotriangle in which pmpm has a small angle. Due to the affine shape of the faces, Δ has an empty interior. We use these triangles as control polygons as shown in **Figure**. 13 and replace the 2-chains by Bézier curves or biarcs (similar to Lemma 2.1). □

In general it is not possible to draw a planar graph pointed using a larger number of (non-crossing) straight-line edges, since every maximal pointed straight-line graph has at most 2n−32n−3 edges [18], anDue to Euler's formula a triangulation has 3n−63n−6 edges. In this sense, Theorem 3.6 is optimal.

We demonstrate the construction used in the proof of Theorem 3.6 by an example. Let G be the graph depicted in **Figure**. 14(a). The big angles of the cpt-assignment and the subdivisions computed by our method are shown in **Figure**. 14(b). This leads to the pointed pseudotriangulation in **Figure**. 14(c) and finally to a pointeDrawing with only three Bézier curves as shown in **Figure**. 14(d).

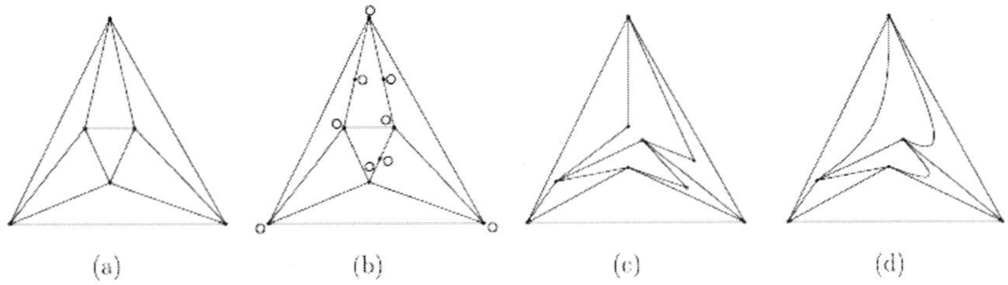

Figure 14. Construction of a pointeDrawing with Bézier curves with help of a combinatorial pseudotriangulation as example.

In the proceedings version of this paper, we made a stronger claim [1, Theorem 7]: for each inner vertex, the face in which it is pointed can be chosen arbitrarily. **Figure**. 15(a) shows a counterexample where this is not true. It is not possible to make the three marked angles big with at most n−3n−3 non-straight edges which are either quadratic Bézier curves or 2-chains. The reason is that a single quadratic Bézier curve bends by less than π. Therefore, in a triangle with three vertices that are pointed in the interior face, all three edges must be curved, see **Figure**. 15(b): if we proceed clockwise along the boundary, the tangent direction turns right by less than π along each edge. At each vertex, it makes a left turn, by pointedness. With less than three curved edges, the tangent direction cannot complete a full right turn of 2π. The same argument works for 2-chains. By a similar argument, a triangle with two vertices that are pointed in the interior face needs at least two curved edges. Applying these facts to our example, we see that all three edges in the triangle ABC must be curved. With a total of n=6n=6 vertices, we have thus exhausted our reservoir of at most n−3=3n−3=3 non-straight edges. But then the two straight edges A′BA′B and A′CA′C together with the curved edge BC cannot make pointed interior angles at B and C in the triangle A′BCA′BC.

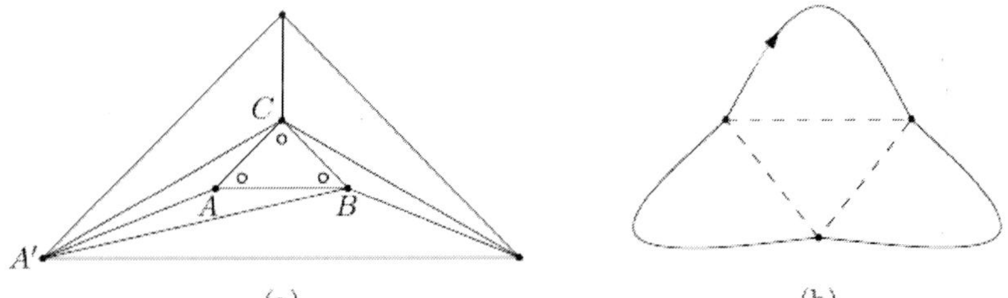

Figure 15. It is not possible to get the three inner vertices pointed in the inner triangle with only 3 quadratic Bézier edges.

This example does not rule out the possibility that pointedness in the chosen faces can be achieved with more than n−3n−3 curved edges, or with biarcs.

ACKNOWLEDGEMENTS

Research on this topic was initiateDuring the fourth European Pseudotriangulation Week in Eindhoven (The Netherlands), organized by Bettina Speckmann. We would like to thank Thomas Hackl, Michael Hoffmann, David Orden, Michel Pocchiola, Jack Snoeyink, and Bettina Speckmann for the inspiring spirit and for many valuable discussions. We also thank the anonymous referees for helpful comments.

REFERENCES

1. O. Aichholzer, G. Rote, A. Schulz, B. Vogtenhuber, PointeDrawings of planar graphs, in: P. Bose (Ed.), Proceedings of the 19th Canadian Conference on Computational Geometry, Ottawa, 2007, pp. 237–240.
2. [2] E.M. Andreev, On convex polyhedra in Lobačevskiĭ space, Math. USSR Sb. 10 (3) (1970) 413–440.
3. G.D. Battista, P. Eades, R.R. Tamassia, I. Tollis, Graph Drawing: Algorithms for Geometric Representations of Graphs, Prentice–Hall, Englewood Cliffs, NJ, 1998.
4. Y. Colin de Verdière, Un principe variationnel pour les empilements de cercles, Invent. Math. 104 (3) (1991) 655–669. [5] C.R. Collins, K.
5. Stephenson, A circle packing algorithm, Comput. Geom. Theory Appl. 25 (3) (2003) 233–256.
6. H. de Fraysseix, J. Pach, R. Pollack, How to draw a planar graph on a grid, Combinatorica 10 (1) (1990) 41–51.
7. I. Fáry, On straight line representations of planar graphs, Acta Univ. Szeged. Sect. Sci. Math. 11 (1948) 229–233.
8. R. Haas, D. Orden, G. Rote, F. Santos, B. Servatius, H. Servatius, D. Souvaine, I. Streinu, W. Whiteley, Planar minimally rigid graphs and pseudo- triangulations, Comput. Geom. Theory Appl. 31 (1–2) (2005) 31–61.
9. P. Koebe, Kontaktprobleme der konformen Abbildung, Ber. Verh. Sächs. Akad. Wiss. Leipzig, Math.-Phys. Kl. 88 (1936) 141–164.
10. G. Laman, On graphs and rigidity of plane skeletal structures, J. Engrg. Math. 4 (4) (Oct. 1970) 331–340.
11. B. Mohar, A polynomial time circle packing algorithm, Discrete Math. 117 (1–3) (1993) 257–263.

12. T. Nishizeki, M.S. Rahman, Planar Graph Drawing, World Scientific, 2004.

13. J. Pach, P.K. Agarwal, Combinatorial Geometry, Wiley–Intersci. Ser. Discrete Math. Optim., John Wiley & Sons Inc., New York, 1995.

14. M. Pocchiola, G. Vegter, Pseudo-triangulations: theory and applications, in: Proceedings of the 12th Annual Symposium on Computational Geometry, SCG '96, ACM, New York, NY, USA, 1996, pp. 291–300.

15. M. Pocchiola, G. Vegter, The visibility complex, Internat. J. Comput. Geom. Appl. 6 (3) (1996) 279–308.

16. G. Rote, F. Santos, I. Streinu, Pseudo-triangulations—a survey, in: Surveys on Discrete and Computational Geometry—Twenty Years Later, in: Contemp. Math., vol. 453, American Mathematical Society, 2008, pp. 343–410.

17. W. Schnyder, Embedding planar graphs on the grid, in: Proceedings of the First Annual ACM-SIAM Symposium on Discrete Algorithms, SODA '90, Society for Industrial and Applied Mathematics, Philadelphia, PA, USA, 1990, pp. 138–148.

18. I. Streinu, Pseudo-triangulations, rigidity and motion planning, Discrete Comput. Geom. 34 (4) (2005) 587–635.

19. H. Whitney, A set of topological invariants for graphs, Amer. J. Math. 55 (1933) 235–321

CHAPTER 6

Extreme Point and Halving Edge Search in Abstract Order Types

Oswin Aichholzer[a, 1], **Tillmann Miltzow**[b], **Alexander Pilz**[a, 2]

[a] *Institute for Software Technology, Graz University of Technology, Austria*

[b] *Institute of Computer Science, Freie Universität Berlin, Germany*

ABSTRACT

Many properties of finite point sets only depend on the relative position of the points, e.g., on the order type of the set. However, many fundamental algorithms in computational geometry rely on coordinate representations. This includes the straightforward algorithms for finding a halving line for a given planar point set, as well as finding a point on the convex hull, both in linear time. In his monograph Axioms and Hulls, Knuth asks whether these problems can be solved in linear time in a more abstract setting, given only the orientation of each point triple, i.e., the set's chirotope, as a source of information. We answer this question in the affirmative. More precisely, we can find a halving line through any given point, as well as the vertices of the convex hull edges that are intersected by the supporting line of any two given points of the set in linear time. We first give a proof for sets realizable in the Euclidean plane and then extend the result to non-realizable abstract order types.

Keywords: LConvex hull; Halving line; Abstract order type; CC System; Chirotope

1. INTRODUCTION

In computational geometry, many fundamental properties of finite point sets do not depend on the actual coordinates of each point in real space, but rather on the relative position of the points among each other. In their landmark paper, Goodman and Pollack [1] capture this idea by defining the order type of a point set. In the plane, two point sets have the same order type if there is a bijection π between the sets s.t. for every triple p,q,rp,q,r of the first set, the corresponding points π(p),π(q)π(p),π(q), and π(r)π(r) have the same orientation (i.e., are both oriented clockwise or counterclockwise).3 This orientation can be tested by the inequality

$$det \begin{pmatrix} p_x & p_y & 1 \\ q_x & q_y & 1 \\ r_x & r_y & 1 \end{pmatrix} > 0,$$

which indicates whether r is to the left of the directed line through p and q , i.e., whether the triple is oriented counterclockwise. The sign of the determinant therefore gives a predicate ∇(p,q,r) that is true iff the triple is oriented counterclockwise. This mapping of all triples of a set to their orientation is also called the chirotope of the set (cf. Remark 1.6 in [2] and [3, p. 95] for details on that term). Many combinatorial properties of a set of points only depend on its order type, like its convex hull, the set of its crossing-free graphs, etc. We implicitly assume throughout this paper that all sets are in general position, i.e., do not contain collinear triples.

In contrast to these properties, there are further, more "metric" properties of a point set that are not determined by the order type. This includes the set's Delaunay triangulation; it is straightforward to construct two sets of the same order type whose Delaunay triangulations are different. Nevertheless, the problem can still be considered as being discrete. Guibas and Stolfi [4] separate topological from geometric aspects, using a predicate InCircle(p,q,r,s)InCircle(p,q,r,s) that is true iff the triple (p,q,r)(p,q,r) is oriented counterclockwise and the point s lies inside the circle defined by the first three points. This predicate is equivalent to

$$det \begin{pmatrix} p_x & p_y & p_x^2 + p_y^2 & 1 \\ q_x & q_y & q_x^2 + q_y^2 & 1 \\ r_x & r_y & r_x^2 + r_y^2 & 1 \\ s_x & s_y & s_x^2 + s_y^2 & 1 \end{pmatrix} > 0.$$

Their Delaunay triangulation algorithm depends almost entirely on this predicate, making it a robust approach, that is intended to be easy to implement and to prove.

Motivated by this approach, Knuth [3] develops axiomatic systems following these two tests. He defines five axioms over a ternary predicate P and calls sets of triples obeying them CC Systems.

Axiom 1 cyclic symmetry. $P(p,q,r) \Rightarrow P(r,p,q).$

Axiom 2 antisymmetry. $P(p,q,r) \Rightarrow \neg P(p,r,q)$.

Axiom 3 nondegeneracy. $P(p,q,r) \vee P(p,r,q)$.

Axiom 4 interiority. $P(t,p,q) \wedge P(t,q,r) \wedge P(t,r,p) \Rightarrow P(p,q,r)$.

Axiom 5 transitivity. $P(p,q,r) \wedge P(p,q,s) \wedge P(p,q,t) \wedge P(p,r,s) \wedge P(p,s,t) \Rightarrow P(p,r,t)$.

These are fulfilled by all point sets in the Euclidean plane, with P=∇ defined as point triple orientation. For these sets, the CC Systems are equivalent to the point set order types. However, there exist CC Systems that cannot be constructed as point sets in R2R2. These are called non-realizable systems, see Section 3.2. CC Systems are equivalent to abstract order types, which, for example, can be used in so-called abstract order type extension [5]. Every abstract order type can be mapped to an arrangement of pseudo-lines in the projective plane. A stretchable arrangement corresponds to a set of realizable order types [3, pp. 34–35]. Goodman and Pollack showed that all CC Systems of up to 8 elements are realizable as point sets [6]. This fact is useful to show properties of small sets by geometric reasoning.

Theorem 1 Goodman, Pollack. Any arrangement of eight pseudo-lines is stretchable.

The concept of the convex hull of a point set generalizes to all CC Systems. The axiomatic approach can also be extended to cover Delaunay triangulations. In the axiomatic settings, Knuth provides $O(n \log n)$ time algorithms for both problems, where the time bound for the latter holds in the expected case. He points out that the algorithm of Guibas and Stolfi uses the coordinate representation to find a line that partitions the point set into two equally sized subsets (cf. [4, pp. 110–111]). Open Problem 1 in [3, pp. 97–98] therefore asks for an algorithm to find such a partition of a CC System in linear time. The problem is straightforward when given an extreme point of the set (i.e., an element of the convex hull boundary). Proving the existence of a linear-time algorithm for finding a single extreme point is also explicitly part of Open Problem 1.

In this work, we answer both parts of the open problem in the affirmative. In Section 2, we give a simple O(n) time algorithm that, given a point c of a set S of size n, finds a halving edge through c ; more specifically, it finds a second point $c' \in S$ s.t. not more than $\lceil \frac{n-2}{2} \rceil$ points are on each side of the supporting line of c and c′. We then describe in Section 3 an algorithm that, given two points p and q, returns the edge of the convex hull that is crossed by the ray from p through q . We first show that the algorithm runs in O(n)O(n) time for realizable sets. We then show that the time bound is also correct for non-realizable sets, that is, for all CC Systems. Both algorithms use a prune and search approach.

Our main motivation is to show that the asymptotic running time for solving these problems does not depend on the representation by coordinates. While an arbitrary halving edge can easily be found by picking a point with median, say, x-coordinate, the problem is more sophisticated when the halving line should pass through a predefined point. E.g., the linear time Ham-Sandwich-Cut algorithm of Lo, Matoušek and Steiger [7] can be adapted to find a halving line through a point. The straightforward way of finding an extreme point of a set given by coordinates is selecting the one with, say, lowest x-coordinate. Finding a convex hull edge that is traversed by a given line in linear time is a subroutine of the so-called Ultimate Convex Hull Algorithm of Kirkpatrick and Seidel [8]. There, the median of the slopes of an arbitrary matching of the points is used for the prune and search approach. Recent algorithms that operate only on triple orientations involve computing the convex hull of disks [9]. Also note that there are models that allow for more abstract order types to be realized than by point sets in the Euclidean plane [10].

A preliminary version of this work appeared as [11].

2. COMPUTING A HALVING EDGE IN LINEAR TIME

In this section we describe an algorithm to compute a halving edge through a given point c of a finite point set S of size n in the plane. While we give the algorithm for realizable order types, note throughout the proof that all assumptions also hold in the abstract setting. The idea is to split the point set evenly by two double wedges centered at c. We identify the double wedge that must contain a halving edge, and continue with the points therein.

We start by picking a point $p_1 \in S \setminus \{c\}$ arbitrarily. Imagine a directed line ℓ rotated by 180° clockwise around c , starting at p1. Let $p_2, \dots, p_{n-1}$ be the remaining points in the order ℓ passes through them, either with its front or its tail. We denote this order by <, which is actually the order described in [3, p. 16]. By ℓ1,…,ℓnℓ1,…,ℓn we denote the "snapshots" of ℓ at the points p1,…,pn−1,p1,…,pn−1,p1 respectively. Thus, ℓ1 and ℓn have opposite directions. See **Figure**. 1 for an illustration.

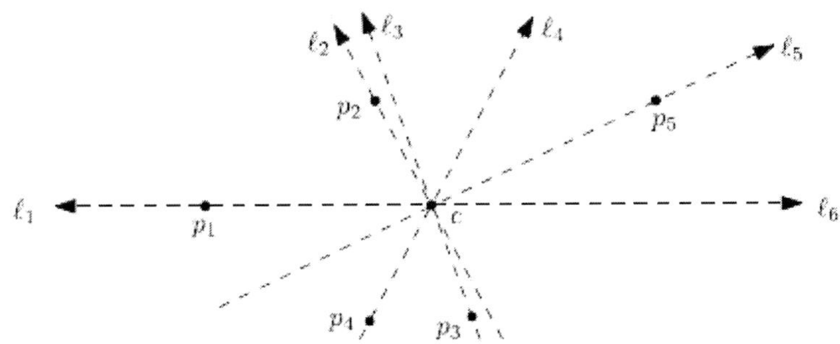

Figure 1. Rotating the line ℓ around c starting at p1 induces an order on the set.

For each pair of $S \setminus \{c\}$, we can compute the relative order in which ℓ passes through them in constant time, using the predicate ∇. Thus the median $m = p_{\lfloor \frac{n}{2} \rfloor}$ with respect to ≺ can be computed in O(n) time [12]. (Note that the number of points to the right of ℓ may not change monotonically while rotating.) The supporting lines of cp1cp1 and cm define two double wedges, containing $S_1 = \{p \in S \setminus \{c\} : p \prec m\}$ and $S_2 = \{p \in S \setminus \{c\} : m \prec p\}$, respectively, $\lceil \frac{n-2}{2} \rceil$ each having not more than $\lceil \frac{n-2}{2} \rceil$ points. See **Figure**. 2. We use the following standard argument. If ℓ1ℓ1 is not a halving line, we have w.l.o.g. more than $\lfloor \frac{n-2}{2} \rfloor$ points to the right of ℓ1ℓ1 and thus less than $\lfloor \frac{n-2}{2} \rfloor$ points to the right of ℓnℓn. Due to our general position assumption, while rotating ℓ only one point changes sides at the same time. Thus, ℓ must be a halving line at some point during the rotation. Assume that $\ell_{\lfloor \frac{n}{2} \rfloor}$ is not a halving line. If less than $\lfloor \frac{n-2}{2} \rfloor$ points are right of $\ell_{\lfloor \frac{n}{2} \rfloor}$, then one of the points in S_1 forms a halving edge with c (recall that there are w.l.o.g. more points right of ℓ1ℓ1). Otherwise one of the points in S2S2 does.

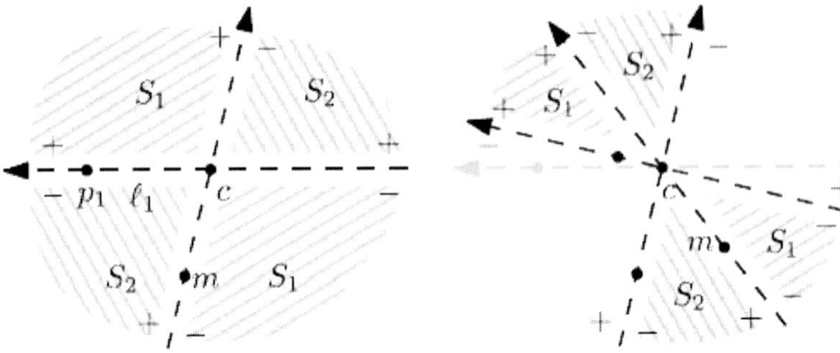

Figure 2. The sign + marks the half plane which contains more $\frac{n-2}{2}$ points. S_1 and S_2 are indicated by the gray tiling patterns. The **Figure**eure shows the first (left) and a later iteration (right).

This way we can decide which subset contains a halving edge, exclude the other subset and iterate. Observe that all the points we exclude belong to two (open) wedges centered at c , one entirely to the left and one entirely to the right of the non-removed lines of $\ell_2, \ldots, \ell_{n-1}$. The closure of each wedge contains a ray of ℓ_1. We store how many points each of these wedges contained. Now, if we want to compute the number of points to the right of some line ℓiℓi in subsequent iterations, then we only need to consider the remaining points and add the number of points belonging to the excluded wedge to the right of ℓiℓi. The algorithm has linear running time, as we exclude approximately half of the points in each linear-time iteration.

Theorem 2. Given a point set $S \subset \mathbb{R}^2$ of size n in general position and a point $c \in S$, , a halving edge of S incident to c can be found in linear time using only the predicate ∇.

Now we argue that the above approach does not require that S is realizable, that is, it works for any CC System (using the predicate P). The order ≺ around c we use is also defined for CC Systems by signing the points in a way that the positive points are to the right of ℓ1ℓ1 and the negative ones are to the left of ℓ1ℓ1, where each negative sign will change the orientation of a triple containing c [3, p. 16] (observe that all our orientation tests include c). This allows us to give a purely combinatorial formulation of our algorithm: Let s be a variable initialized to 0 and let riri be the number of negative points preceding point pipi plus the number of positive points succeeding pipi with respect to ≺ in the current iteration (this corresponds to the points to the right of ℓiℓi). We may assume that initially we have $r_1 > \frac{n-2}{2} > r_n$. If the sum of rmrm and s is less than $\lfloor \frac{n-2}{2} \rfloor$, we remove the points that succeed m, and add the number of removed positive points to s. Otherwise, we remove the points that precede m and add the number of removed negative points to s . Note that for every point pipi, the value of ri+sri+s is an invariant over all iterations. Eventually, we end up with an element pjpj such $r_j + s = \lfloor \frac{n-2}{2} \rfloor$.

In the combinatorial formulation of the algorithm, the crucial properties of the sequence r we use is that subsequent elements change by one and that the required value is between r1r1 and rnrn. Therefore, the algorithm is also adaptable to more sophisticated problems than finding a halving edge. One example is the search for a balanced line, that is, a line that separates a set consisting of n white points and n black points in a way that the difference between the black and white points on each side of the line is 0 (additional care has to be taken when defining the sequence r since a balanced line does in general not pass directly through a given point c, i.e., there is no explicit representation of the line as a supporting line of two points). An application of this can be found in [13]. There, it is shown that certain crossing-free matchings on a point set are not unique by constructing a balanced line crossing a segment of a given matching.

For realizable sets, the algorithm can directly be extended to higher dimensions. Order types in RdRd are defined analogously as in the plane, see [1]. The orientation of a (d+1)(d+1)-simplex can be computed by the corresponding determinant. Let (p1,…,pd+1)(p1,…,pd+1) be an ordered tuple of points in RdRd. For the oriented hyperplane H through the points (p1,…,pd)(p1,…,pd), ∇(p1,…,pd+1)∇(p1,…,pd+1) indicates on which side of H the point pd+1pd+1 lies.

Corollary 3. Given a point set S⊂RdS⊂Rdin general position, d constant, anDistinct points c1,…,cd−1∈Sc1,…,cd−1∈S, one can find a halving hyperplane through c1,…,cd−1c1,…,cd−1in linear time using only ∇.

3. COMPUTING A CONVEX HULL EDGE IN LINEAR TIME

In this section we show how to compute a convex hull edge of a set in linear time, only using the orientation of triples. We first present the algorithm for point sets and then extend it to general CC Systems.

3.1. Realizable point sets

As a first step we describe an algorithm called BasicMin, which plays a crucial role as a subroutine. Let S be a point set in the plane and suppose we are given two points $p, r \in S$. . We assume that pr is a halving edge of S and that n=|S| is even. W.l.o.g. let r be the coordinate origin and let p be on the positive part of the x-axis. Let M be an arbitrary perfect matching between the points above and below the x -axis, i.e., for any edge $s = ab \in M$ we have $\nabla(p, r, a) \neq \nabla(p, r, b)$. See **Figure**. 3 for an illustration.

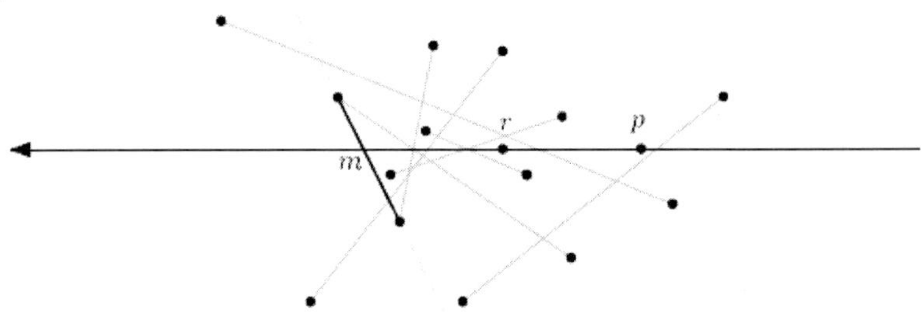

Figure 3. An arbitrary matching of edges that partition S, and m, the result of BasicMin.

Let $\ltimes$ be the binary operator that accepts two edges $s, s' \in M$ as input and returns the edge on the convex hull boundary of $s \cup s'$ that crosses the x-axis at the smallest x-coordinate, i.e., the pair of endpoints whose upward-directed supporting line has all other points of s and s' to the right. The relevant property of the operator is that the crossing of $\ltimes(s, s')$ with the x-axis is not to the right of the crossings of s and s's' with the x-axis.

BasicMin takes a point set S and two points p and r as input, partitions S arbitrarily into the matching M= $\{s_1, \ldots, s_{(\frac{n-2}{2})}\}$, and computes a special edge $m = m_{(\frac{n-2}{2})}$ iteratively via

$$m_1 = s_1,$$

$$m_{(i+1)} = \ltimes(m_i, s_{(i+1)}).$$

Obviously, the running time of BasicMin is linear in n. Note that m does not need to be on the convex hull of the whole set. Also, m may depend on the (undefined) order in which the elements of M are processed by BasicMin. However, m has the following useful property.

Lemma 4. *Let ℓ be a line directed upwards, crossing the x-axis left of the crossing of m with the x-axis. Then at least $n-2$ points of S are to the right of ℓ*

Proof. Note that the crossing of ℓ with the x-axis is further to the left than any other such crossing of M. Assume, for the purpose of contradiction, that there is an edge in M for which both points lie to the left of ℓ. Then also the crossing with the x-axis is to the left of the crossing of m, a contradiction. Hence, for each of the $n-2$ edges in M at least one endpoint lies to the right of ℓ.

Note that, even though the argumentation of Lemma 4 involves relative positions of crossings, the constant-time opera- tion ⋉ can be expressed by using only ∇. Now we are ready to give the main algorithm.

Theorem 5. *Given two points p, q of a point set $S \subset \mathbb{R}^2$ of size n in general position, one can find the edge e of the set's convex hull that passes through the ray pq in $O(n)$ time using the predicate* ∇.

Proof. W.l.o.g. let pq be horizontal with q left of p. Note first that the case where q is a vertex of the convex hull can be identified in linear time. We therefore concentrate on the setting where one endpoint of e is below pq and the other one is above. Let U be the set strictly above pq, whereas L is the set strictly below pq. W.l.o.g. let $|U| \geqslant \frac{n-2}{2}$.

Consider the endpoint $u1 \in U$ of e. If we remove $u1$, the ray pq intersects the boundary of the (new) convex hull at a new edge e' with an endpoint $u2 \in U$. Note that the other endpoint of e' might now be q. In any case, iteratively removing points from U of the intersected convex hull edge induces an order on U (see **Figure**. 4). Note that this corresponds to the order in which the points of U are traversed by a tangent t of $\mathrm{CH}(L \cup \{p,q\})$ that is rotated clockwise around that hull,

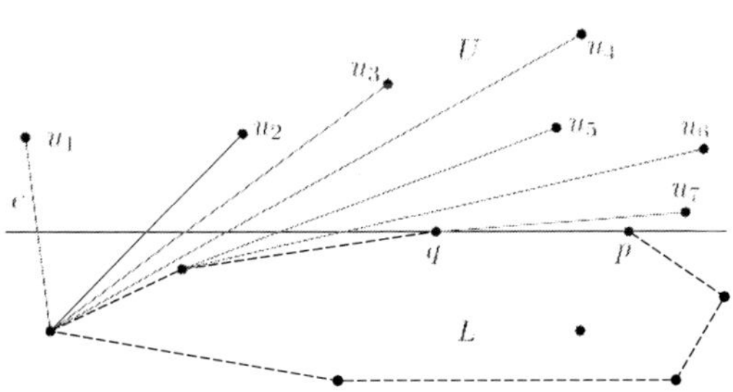

Figure 4. The order on U defined by removing vertices of the intersected convex hull edge.

starting at e and ending at pq. The main observation is that in the search for $u1$, given a point ui and the tangent t passing through ui, we can discard all points of U to the right of ui with respect to the point l where t touches $\mathrm{CH}(L \cup \{p, q\})$, since none of these points can be $u1$. The support $l \in L \cup \{q\}$ of t can be found in linear time, since the radial order of $L \cup \{p, q\}$ around any ui is linear.

Note that these observations already imply the following randomized approach. Select any element uiui of U at random. The other support l of the tangent (recall that this might as well be q) can be found in linear time. We discard the points of U right of luilui and iterate. However, consecutive "bad" choices of uiui result in overall quadratic worst-case behavior. We therefore have to make a "good" choice of uiui in order to discard a linear number of points per iteration.

The points of U are ordered linearly around p. Let ur be the median of this order, which we select in linear time. Let M be an arbitrary perfect matching between the points of U to the left and to the right of pur (maybe omitting one point), see **Figure**. 5. Now we apply BasicMin on M with r = ur , which results in an edge m = uaub . By construction, all edges of M as well as uaub cross the ray pur . Now, find the tangents ta and tb of CH(L ∪ {p, q}) through ua and ub , respectively (we consider them being directed upwards). Let ℓ = ta if ub is to the right of ta , otherwise let ℓ = tb . Note that p is right of ℓ since q is included in the set that we place the tangent on. There are two cases to consider. If ℓ crosses the ray pur at a point x, then the crossing of uaub is on the ray between p and x (by definition of ℓ). Due to Lemma 4, at least half of the points of M are to the right of ℓ. Otherwise, if ℓ does not cross the ray pur , then all points of U to the right of the ray are also to the right of ℓ. In both cases, we can discard at least half of the points, which is at least a quarter of the overall set S (recall that U was w.l.o.g. larger than L; in each iteration, the process is applied to the larger of the two sets). We can therefore in linear time reduce this problem to constant size such that it then can be solved by a brute-force approach.

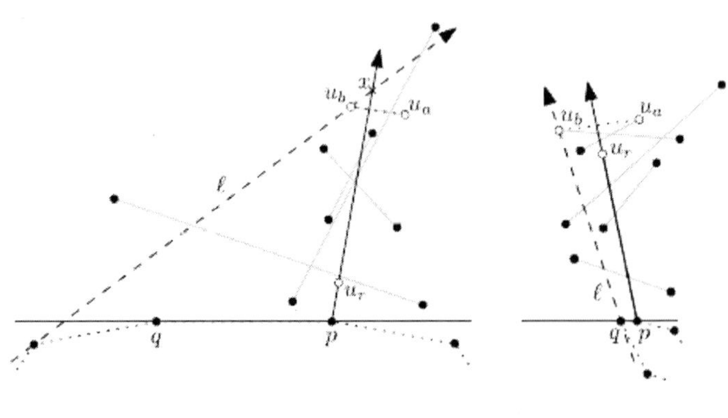

Figure 5. The edge uaubuaub allows to prune half of the upper points.

Note that the transitivity of the order on U directly follows from the definition of the convex hull and carries over to general CC Systems. Also, the transitivity of LU{p,q}LU{p,q} around any point of U holds for CC Systems due to Axiom 5. This already implies that the algorithm is correct for any CC System. However, the linear time bounDepends on the number of points that are to the right of ℓ. In order to show that the bound also holds for abstract order types, we need to prove that BasicMin also works as expected on non-realizable sets, and that then also ℓ has at least half of the points of U to the right.

3.2. Non-realizable sets

There exist several equivalent definitions for the convex hull of a finite point set S in the plane. For example, the convex hull can be defined as the convex polygon of smallest area that contains all points of the set. Equivalently, the boundary of the convex hull of S is the sequence of pairs uv with $\{u,v\} \subset S$ such that all points $S \setminus \{u,v\}$ are to the left of uv. This latter definition is more combinatorial and carries over to the abstract setting, where we are no longer concerned with "geometric artefacts" like area or perimeter.[4] In this subsection, we will give a brief summary of the relevant properties of abstract order types, following Goodman and Pollack [2], which we suggest to the reader for further information on that topic, in addition to [3]. Throughout this section, see **Figure**. 6 for illustrations.

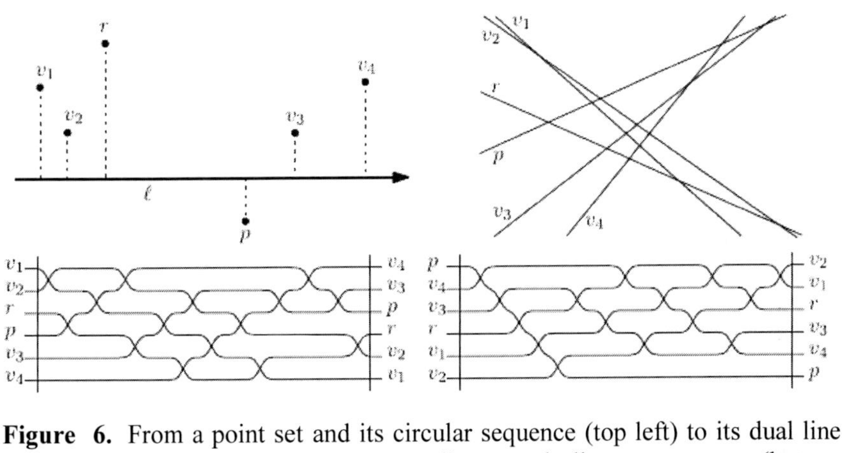

Figure 6. From a point set and its circular sequence (top left) to its dual line arrangement (top right) and a corresponding pseudo-line arrangement (bottom left). The transformed arrangement to the bottom right represents the same order type.

Let us again start in the Euclidean plane by considering a point set S, which we still require, for ease of presentation, to be in general position (see [2] on how to handle other cases). Let ℓ be a directed line not orthogonal to any supporting line of S. Project the points orthogonally on ℓ, which gives us a sequence of the points along ℓ. If we rotate ℓ in counterclockwise direction, then eventually two points, let them be u and v, will happen to be projected to the same point on ℓ. When continuing the rotation process, u and v will have changed their position in the sequence along ℓ. After a rotation of 180°, the initial sequence of the projected points on ℓ will be reversed. Continuing the rotation gives an infinite periodic sequence of transpositions like the one of u and v. It is called the circular sequence of permutations associated to S, or the circular sequence of S, for short. One can observe that after a reversal of u and v all other pairs are reversed until u and v are reversed again.

The circular sequence allows us to observe several geometric properties of S. A point that precedes all others in some permutation (on ℓ) in the circular sequence is an extreme point. Note that all the information we can get from the circular sequence is encoded in one half-period. We therefore can identify the vertices of the convex hull as those points that are topmost or bottommost on ℓ at some point, when we look at ℓ such that it is directeDownwards. For every subset of S, the circular sequence therefore gives the point on its convex hull. Consider again the moment when u and v change their position in the permutation, say from uv to vu. Then the triple uvw is oriented counterclockwise if and only if w is projected below (i.e., after) both u and v when they change their position (assuming counterclockwise rotation of ℓ). Note that this requires that at the beginning of the half-period u is above (before) v. It is common and convenient to identify the points of S with their order at the initial position of ℓ . For S={p1,...,pn}S={p1,...,pn}, we therefore start with pipi above pjpj on ℓ iff i<ji<j. While the starting position of ℓ in the half-period is arbitrary, this labeling is useful when comparing point triples. After a half-period, the points are given in the sequence ⟨pn,...,p1⟩ ⟨pn,...,p1⟩ .

The circular sequence contains strictly more information than the order type of a set. While the order type is implied by the circular sequence, it is rather straightforward to construct point sets with different circular sequences that have the same order type.

Consider an arrangement of pseudo-lines in the projective plane P2 (a set of closed curves not separating P2 that pairwise intersect in exactly one point). An arrangement is simple if no three pseudo-lines intersect in one point. One important property of pseudo-line arrangements, shown by Levi [14], is that they can be extended by additional pseudo-lines.

Lemma 6 Levi Enlargement Lemma.

Given a pseudo-line arrangement AAin P2and two points that do not both lie on the same element of AA, there exists a pseudo-line arrangement AU{χ}AU{χ}such that the pseudo-line χ passes through these two points.

Using this lemma, choose a pseudo-line ℓ∞ℓ∞ not part of the arrangement as the line at infinity. Place a point ψ on ℓ∞ℓ∞. If we choose ψ as the point at vertical infinity, we can identify the arrangement of pseudo-lines with an arrangement of x-monotone pseudo-lines in the Euclidean plane (where the curves are no longer closed). This requires the application of the Levi Enlargement Lemma to get x-monotonicity, details are given in [2]. If we sweep this arrangement from left to right, we get again a half-period of a circular sequence, where a change of a pair happens as we sweep across the intersection of two pseudo-lines. Goodman and Pollack [2] show that any circular sequence of a point set describes the topology of an arrangement of pseudo-lines in P2. Actually, for every point set in the Euclidean plane, the standarDuality transform for point sets gives an arrangement of straight lines whose crossings, when sweeping from left to right, give a half-period of the circular sequence of the set [15]. There are, however, arrangements of pseudo-lines in the projective plane that are not stretchable. For them, no point set in the Euclidean plane exists that implies any sequence derived from such an arrangements by any choice of ψ.

Recall that we were able to derive the orientation of a point triple uvw by the position of the points in a half-period of the circular sequence. Given a direction around a point at infinity allows us to give a triple of pseudo-lines an orientation, even if there does not exist a corresponding point set order type in the Euclidean plane (i.e., the sequence is non-realizable); we still get a predicate ∇a∇a that gives each triple an orientation, that is, an abstract order type. Knuth [3] shows that, for any simple arrangement, ∇a∇a fulfills the five axioms defining CC Systems, and from any CC System a corresponding arrangement can be derived; hence, the terms "CC System" and "abstract order type" are merely synonyms. We can use either the axioms or a pseudo-line arrangement (together with a direction around a point at infinity) for arguing about abstract order types. In the next section, we will use the model of x-monotone pseudo-lines in the Euclidean plane.

At first sight, the identification of abstract order types with pseudo-line arrangements in the projective plane might feel like a detour to x-monotone arrangements in the Euclidean plane as representation of circular sequences. This might be the case for the application of the concept in this paper, however, it is profound in the whole theory about abstract order types. Recall that different circular sequences might correspond to two point sets having the same order type. This is also the case for non-stretchable arrangements. (On the other hand, for arrangements of pseudo-lines in the projective plane, there are transformations that are order-type-preserving and some that are not.) If two pairs uv and st of four different points change their position in the sequence and uv happens right before st in the half-period, then changing st right before uv results in the same order type. (To reduce the cases to consider, we will nevertheless make use of that relative position of pseudo-line crossings in the next section.) There is another aspect that is not directly captured by x-monotone arrangements in the Euclidean plane. When mapping the arrangement from the projective to the Euclidean plane, we chose the line at infinity through ψ arbitrarily; however, the topology of the arrangement in the projective plane does not depend on the choice of this line. Analogously, the circular sequence of a point set does not depend on the actual starting position of ℓ. In terms of pseudo-lines in the Euclidean plane, the following can be shown (more details are given in [2] and [3]). Consider a crossing of two pseudo-lines that has no further crossings to the left from its initial position. Then we can move this crossing to the right end of the arrangement, such that the resulting arrangement still represents the same abstract order type. In other words, we "untangle" the crossing at the left end and introduce a new one to the right (note that this corresponds to removing the position change uv at the beginning by a change vu at the end). For every abstract order type and any crossing therein, we can therefore choose a x-monotone representation where that crossing is leftmost.

3.3. A general proof of the time bound

To see why care has to be taken when we deal with non-realizable order types, note that in general the order type does not capture the relative position of the supporting lines of point pairs and the crossings of these lines. However, abstract order types capture some of the information that is related to crossings of supporting lines, which allows us to show that the properties needed for BasicMin are also present in the abstract setting.5 We use the dual representation of abstract order types in the Euclidean plane by x -monotone pseudo-lines; this allows us to use the obvious meaning of terms like "above" and "below" when describing an arrangement AA (e.g., a pseudo-line a is above a point b when the point on a that has the same x-coordinate as b has a larger y-coordinate than b).

Let us recall the problem setting. We are given a set S of n elements (which we call points, even though S might not be a realizable point set), containing two special points p and q. The set S is separated by pq into a set U to the right of pq, and a set L to the left of it (where "left" and "right" are indicated by a predicate ∇_a). We want to obtain the pair $l_1 u_1$, $l_1 \in L, u_1 \in U$ that is consecutive on the convex hull of S, where p is to the right of $l_1 u_1$. This is done by obtaining a pair $\ell = l_j u_i$ such that at least half of the points of U (minus a constant) are to the right of ℓ and no point of L is to the left of ℓ..

Consider the dual x-monotone pseudo-line arrangement A representing the abstract order type of S. Keep in mind that p is an extreme point of the set $U \cup \{p\}$, and that $r = u_r$ is the median of the points in U ordered radially around p. We can represent the arrangement such that the crossing of the pseudo-lines p and r (which corresponds to the supporting line of pr in the primal[6]) is the leftmost crossing in the x-monotone representation of A. The linear order of the points around p in the primal splits the set $U \setminus \{u_r\}$ into *left* and *right* points, separated by pr. In the dual, the right pseudo-lines pass above the crossing pr, and the left pseudo-lines pass below. Recall the description of BasicMin. M is an arbitrary perfect matching between the left and the right points. The operator ⋉ accepts two point pairs, each pair consisting of a left and a right

point. The output of the operator is a pair *z* consisting of a left and a right point such that all other points are to the right of the oriented line through these points. This pair *z* is well-defined in the abstract setting as well (and there is always a geometric representation due to Theorem 1). Recall that we compute a special pair $m = m_{\left(\frac{n-2}{2}\right)}$ iteratively via

$$m_1 = s_1,$$

$$m_{(i+1)} = \ltimes(m_i, s_{(i+1)}).$$

The crucial property of the pair m=uaubm=uaub (and later the line ℓ tangent to CH(L∪{p,q})CH(L∪{p,q}) through uaua or ubub) is that at least one endpoint of each pair si∈Msi∈M lies on the same side of m as the pivot p. (We assume that m and the elements of M are directed from the left point to the right point and hence p is to the right of m.) In the dual, each element of M is represented as the crossing of a right and a left pseudo-line in AA (this corresponds to the supporting line of the matched pair of points in the primal). The crucial property for m in the dual therefore is to have at least one pseudo-line of each pair si∈Msi∈M passing below its crossing in AA. For realizable point sets, we were able to argue for the correctness of BasicMin using the intersection χ of a matched pair s with the supporting line of p and r. In the dual, this intersection χ corresponds again to a pseudo-line that can be added to AA; this pseudo-line passes through the crossing pr and s. Also for non-realizable sets, such a pseudo-line χ exists due to the Levi Enlargement Lemma. In fact, in non-realizable settings the intersections behave in the same transitive manner as in the realizable setting; see **Figure**. 7 for an illustration of the following statement.

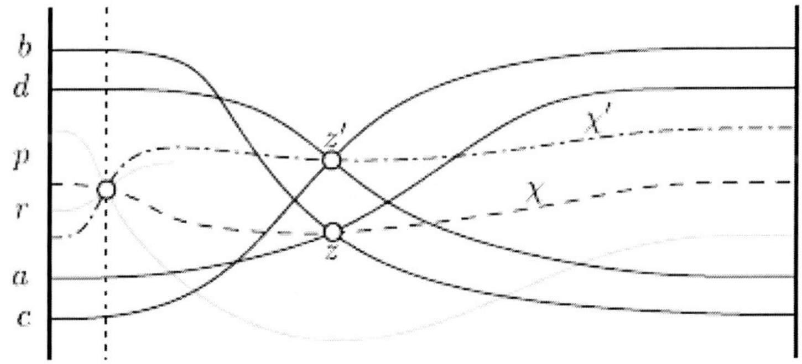

Figure 7. Observation 7: A pseudo-line χ′χ′ (not part of the arrangement) witnesses that no element is left of z′z′ in the primal.

Observation 7. Let z=(a,b)z=(a,b)and z′=(c,d)z′=(c,d)be two pairs such that the point p is to the right of both pairs (i.e., in the dual the pseudo-line p is below the crossings z and z′z′). Let χ (χ′χ′) be a dual pseudo-line through the crossings pr and z (z′z′). If none of a and b is to the left of the primal line z′z′, then the dual pseudo-line χ′χ′is above χ in the part of the dual arrangement that is to the right of the dual crossing pr. This also holds if $a = c \text{ or } b = d.$.

Hence, after applying BasicMin to the matching M, we obtain in the dual a pseudo-line crossing m on a pseudo-line χmχm that passes through the crossing pr, such that no crossing of M is above χmχm. Suppose, for the sake of contradiction, that there is a pair (a,b) such that both dual pseudo-lines a and b pass above the crossing m, but the crossing ab is below χmχm. If the crossing ab is to the left of m in the x-monotone arrangement, then, when traversing a from left to right, one would have to pass below χmχm and then go above it again before m. Otherwise, if the crossing ab is to the right of m, then b would have to intersect χmχm to be above it at m and then has to be below χmχm again to reach the crossing ab. Both cases contradict the fact that a pair of pseudo-lines intersects exactly once in the x-monotone arrangement. Hence, half of the pseudo-lines are below the pseudo-line crossing m in AA. This corresponds to at least half of the points being on the same side of m as p.

Recall that in the proof of Theorem 5, we chose a directed line ℓ supported by a point l∈L∪{q}l∈L∪{q} and either uaua or ubub such that no elements of L∪{p,q,ua,ub}L∪{p,q,ua,ub} is to the left of it. We now proceed to show that at least one point of each pair (v,w)(v,w) in M is to the right of the directed line ℓ, as demanded in the proof of Theorem 5. The points involved are {p,q,r,l,ua,ub,v,w}{p,q,r,l,ua,ub,v,w}, where l and q might be the same point. Since these are at most eight points, we are allowed to use geometric arguments due to Theorem 1. However, we must not rely on the positions of the crossing points on the ray pr. Suppose, for the sake of contradiction, that neither of v and w is right of ℓ. At least one of v or w has to be to the right of uaubuaub. The line ℓ separates v and w from the remaining subset. Further, v and w are separated by pr. These observations imply that vw is an edge of the convex hull of {p,ua,ub,v,w}{p,ua,ub,v,w}. However, this means that the crossing of the pseudo-lines v and w is above uaua and ubub in AA, which contradicts the fact that uaubuaub is m. Thus, we conclude

Theorem 8. Theorem 5also holds for non-realizable CC Systems, i.e., abstract order types.

4. CONCLUSION

We presented two algorithms that only use the information whether a point triple is oriented clockwise or counterclockwise. Both, a halving edge through a given point and a convex hull edge crossing a specified ray, can be found in linear time, without being given the coordinate representation. We showed that the algorithms also work for general CC Systems (i.e., abstract order types), and thus answer a long-standing open problem of Knuth [3] in the affirmative.

Note that the parts of the so-called Ultimate Convex Hull Algorithm by Kirkpatrick and Seidel [8] that depend on coordinates are essentially the one that find the convex hull edge on the ray that separates a subproblem into two parts. Also, Chan's output-sensitive algorithm [16] can be implemented in our setting using Theorem 5. Both allow to improve the time bound given in [3] for realizable point sets regarding output-sensitivity to $O(n\log h)$<img height="13" border="0" style="vertical-align:bottom" width="58" alt="View the MathML source" title="View the MathML source" src="http://origin-ars.els-cdn.com/content/image/1-s2.0-S0925772113000394-si212.gif">O(nlogh) for h extreme points.

ACKNOWLEDGEMENTS

We thank Thomas Hackl, and Günter Rote, as well as anonymous referees, for helpful suggestions on improving the presentation of the results. In particular, one of the referee's remarks allowed to significantly shorten the correctness proof for the extreme point algorithm in the abstract setting.

REFERENCES

1. J.E.Goodman,R.Pollack,Multidimensionalsorting,SIAMJ.Comput.12(1983) 484–507.
2. J.E.Goodman,R.Pollack,Semispacesofcon**Figure**urations,cellcomplexesofarrangements,J.Combin.TheorySer.A37(1984)257–293.
3. D.E.Knuth,AxiomsandHulls,LectureNotesComput.Sci.,vol.606,Springer,1992.
4. L.J.Guibas,J.Stolfi,Primitivesforthemanipulationofgeneralsubdivisionsandcomputationof Voronoidiagrams,ACMTrans.Graph.4(1985)74–123.
5. O.Aichholzer,H.Krasser,Abstractordertypeextensionandnewresultsontherectilinearcrossingnumber,Comput.Geom.36(2007)2–15.
6. J.E.Goodman,R.Pollack,ProofofGrünbaum'sconjectureonthestretchabilityofcertainarrangementsofpseudolines,J.Combin.TheorySer.A29(1980)385–390.
7. C.-Y.Lo,J.Matoušek,W.Steiger,Algorithmsforham-sandwichcuts,DiscreteComput.Geom.11(1994)433–452.
8. D.G.Kirkpatrick,R.Seidel,Theultimateplanarconvexhullalgorithm?SIAMJ.Comput.15(1986)287–299.

9. L.Habert,M.Pocchiola,Computingtheconvexhullofdiscsusingonlytheirchirot ope,in:Proc.20thEuropeanWorkshopComput.Geom.(EuroCG2004),Seville, Spain,2004,pp.111–114.

10. O.Aichholzer,M.Korman,A.Pilz,B.Vogtenhuber,Geodesicordertypes,in:J.G udmundsson,J.Mestre,T.Viglas(Eds.),Proc.18thInternationalComputingand CombinatoricsConference(COCOON2012),in:LectureNotesComput.Sci.,vo l.7434,Springer,2012,pp.216–227.

11. T.Miltzow,A.Pilz,Selectionofextremepointsandhalvingedgesofasetbyitschir otope,in:Proc.28thEuropeanWorkshopComput.Geom.(EuroCG2012),Assisi, Italy,2012,pp.85–88.

12. M.Blum,R.W.Floyd,V.Pratt,R.L.Rivest,R.E.Tarjan,Timeboundsforselection ,J.Comput.SystemSci.7(1973)448–461.

13. A.Asinowski,T.Miltzow,G.Rote,Quasi-parallelsegmentsandcharacterizationofuniquebichromaticmatchings,in:Proc. 29thEuropeanWorkshopComput.Geom.(EuroCG2013),Braunschweig,Germ any,2013,pp.225–228.

14. F.Levi,DieTeilungderprojektivenEbenedurchGeradeoderPseudogerade,Ber. Math.-Phys.Kl.Sächs.Akad.Wiss.Leipzig78(1926)256–267,inGerman.

15. J.E.Goodman,R.Pollack,Atheoremoforderedduality,Geom.Dedicata12(1982)63–74.

16. T.M.Chan,Optimaloutput-sensitiveconvexhullalgorithmsintwoandthreedimensions,DiscreteComput.G eom.16(1996)361–368.

CHAPTER 7

Fractional and *j*-Fold Coloring of the Plane

Jarosław Grytczuk _[1,2], Konstanty Junosza-Szaniawski [2], Joanna Sokół[2], and Krzysztof Wesek[2]

[1]*Faculty of Mathematics and Computer Science, Jagiellonian University, Poland,*

[2] *Faculty of Mathematics and Information Science, Warsaw University of Technology, Poland,*

ABSTRACT

We present results referring to the Hadwiger–Nelson problem which asks for the minimum number of colors needed to color the plane with no two points at distance 1 having the same color. Exoo considered a more general problem concerning graphs $G_{[a,b]}$ with $\mathbb{R}^2$ as the vertex set and two vertices adjacent if their distance is in the interval $[a,b]$. Exoo conjectured $\chi(G_{[a,b]}) = 7$ for sufficiently small but positive difference between *a* and *b*. We partially answer this conjecture by proving that $\chi(G_{[a,b]}) \geq 5$ for $b > a$. . A *j*-fold coloring of a graph $G = (V, E)$ is an assignment of *j*-elemental sets of colors to the vertices of *G*, in such a way that the sets assigned to any two adjacent vertices are disjoint. The fractional chromatic number $\chi_f(G)$ is the infimum of fractions *k* / *j* for *j*-fold coloring of *G* using *k* colors. We generalize a method by Hochberg and O'Donnel (who proved that $G_{[1,1]} \leq 4.36)$ for the fractional coloring of graphs $G_{[a,b]}$, obtaining a bounDependent on *ab* ab. We also present few specific and two general methods for *j*-fold coloring of $G_{[a,b]}$, for small *j*, in particular for $G_{[1,1]}$ and $G_{[1,2]}$. The *j*-fold coloring for small *j* has strong practical motivation especially in scheduling theory, while graph $G_{[1,2]}$ is often used to model hidden conflicts in radio networks.

Keywords: Fractional coloring Hadwiger–Nelson problem Coloring of the plane

1 INTRODUCTION

1.1 Mathematical Context

The famous Hadwiger–Nelson problem asks for the minimum number of colors required to color the Euclidean plane $\mathbb{R}^2$ in such a way that no two points at distance 1 from each other have the same color. The question can be equivalently stated in the graph theory language: Recall that a *coloring* of a graph $G = (V, E)$ is a function $c : V \to K$ (where K (where *K* is a set of colors) such that every two adjacent vertices *x*, *y* satisfy $c(x) \neq c(y)$. The *chromatic number* $\chi(G)$ of *G* is the minimal cardinality of the set of colors to color *G*. Therefore, the Hadwiger–Nelson problem is about determining the chromatic number of the graph on the set of vertices R 2 R2 with vertices in distance one adjacent—it is called *unit distance graph* and in this article will be denoted by $G_{[1,1]}$ (the notation will be justified shortly).

The problem was originally proposed by Edward Nelson in 1950 and was made publicly known by Hugo Hadwiger [4]. Pioneers on the topic observed the following bounds: Nelson first showed that at least 4 colors are needed (see the proof by Moser and Moser [7] which uses the so-called Moser spindle) and John Isbell was first to prove that 7 colors are enough (this result was published by Hadwiger [4]). Somehow surprisingly, the aforementioned bounds remain unchanged since their origin in 1950s, i.e. for more than 60 years nobody has found anything sharper than $4 \leq \chi(G) \leq 7$ - as long as we consider the full generality. Nevertheless, advanced studies on the question, its subproblems and other related topics provide some understanding. For example, if we consider only measurable colorings (i.e. with measurable colors) then at least 5 colors are necessary [2] and if coloring of the plane consists of regions bounded by Jordan curves then at least 6 colors are required [12]. Generally, across the decades the Hadwiger–Nelson problem inspired many interesting results in the touchpoint of combinatorics and geometry, a vast number of challenging problems and various applications [10].

One of the possible ways to generalize the first question was presented by Exoo [1] (see also [6]). He considered graphs on the set of vertices R 2 R2 with vertices in distance in the interval [*a*, *b*], we denote such graphs by

$G[a,b]$ G[a,b]. How many colors would be enough to color such a graph, depending on a and b? Are there some important ranges of those parameters?

The other path for research leading from the Hadwiger–Nelson problem concerns different models of graphs coloring. In the majority of this article, we investigate fractional colorings, in some sense a generalization of the classic coloring: every vertex gets a j-elemental set of colors from the set of colors (of size k) and the sets for adjacent vertices have to be disjoint. The 'quality' of such a coloring is measured by the fraction k / j and the fractional chromatic number is the infimum of such fractions (one can also ask for the best fraction with a fixed j). It can be seen as in this case every single color is just a $1 / j$-size part of a 'complete color', so we divide 'complete colors' from the classic coloring to somehow save a bit by combining partitioned colors in a clever way. It turns out that in fact we can save much using fractional coloring: fractional chromatic number is always lower or equal to chromatic number of a given graph, but the difference can be arbitrarily large.

The fractional chromatic number of the graph $G_{[1,1]}$ has been studied in the literature. The best upper bound is due to Hochbeg and O'Donnell [5] (based on an idea by Fisher and Ullman [3] of looking for a dense subset of the plane which avoids unit distance) and the best lower bound (using a finite subgraph of the plane) can be found in the book of Scheinerman and Ullman [9] (along with a good explanation of the upper bound): $3.555 \leq \chi_f(G_{[1,1]}) \leq 4.36$. Note that the upper bound is much smaller than the upper bound for $\chi(G_{[1,1]})$.

1.2 Practical Motivation

Coloring of such geometrical graphs has also some significant practical motivations in telecommunication. We will briefly describe an example. More on this topic can be found in the paper of Walczak and Wojciechowski [11].

Consider the following problem: We are given a set of transmitters with equal ranges placed in some area (assume that ranges are equal to 1)—some of them are in each other range, and some of them are not. If two transmitters are in each other's range, we assume that they can 'quickly' agree on their communication (there are algorithms for it). If two transmitters are not in each other's range but have a common neighbor C, then it is possible that A and B would try to transmit to C in the same time—in this case C cannot listen to both messages. Hence we have a conflict which cannot be solved by a direct communication between A and B. If two transmitters A and B are not in each other's range anDo not have a common neighbor, then there is no conflict. The problem is to assign time-slots for transmitters (in an equitable way) such that no two conflicted transmitters share a time interval.

How can we use graph coloring in this problem? We can create the graph G of conflicts for this network of transmitters: vertices correspond to transmitters and two vertices are adjacent if the corresponding transmitters X, Y are at distance greater than 1 and have a common transmitter in their respective ranges (note that it is possible only if X and Y are at distance in the set (1, 2)). Hence, using a k-coloring of G, such that each color corresponds to one time-slot is one of the possible ways of constructing a proper transmission-schedule. By the definition of coloring, the produced schedule does not contain any pair of conflicted transmitters sharing their time of transmission and every transmitter gets the same amount of time in one cycle of transmission. The length of the schedule is k. On the other hand, we can make use of a fractional coloring of G. Since we demand that in one cycle of transmission every transmitter gets a unit of time, then every color in a fractional coloring corresponds to an interval of length $1/j$ of the time unit. This model guarantees avoiding of conflicts as well as equability of the schedule (if we used k colors then the length of the schedule is k/j). The relation between considered coloring models implies that fractional coloring can possibly produce a shorter schedule in comparison to classic coloring, which (in real world) means: we can save a bit of time. However, the technical constraints imply that too big values of j are not accepted—too fragmented schedule is not practical. This issue plays an important role in our work—we devote a part of this article to fractional colorings with 'small j'.

Now, consider a graph on the set of vertices $\mathbb{R}^2$ with vertices at distance in the set [1, 2] adjacent, which will be denoted in this article as $G_{[1,2]}$. Clearly, every conflict graph of a transmitters network as described above is a subgraph of $G_{[1,2]}$. Therefore, any coloring or fractional coloring of infinite $G_{[1,2]}$ induce, respectively, a coloring or a fractional coloring of a given finite conflict graph. Hence, we get a universal scheme for any network. Additionally, this universal scheme works also if the transmitters are placed on moving vehicles (thus changing the conflict graph). This is an indisputable advantage over using standard coloring algorithms for a given conflict graph.

2 PRELIMINARIES

First, we need to formally define fractional and classic coloring of graphs.

Definition 1 A *coloring* of a graph $G = (V, E)$ with k colors (or k-coloring) is an assignment of colors $\{1, 2, ..., k\}$ to the vertices of G such that no two adjacent vertices have the same color. The smallest number of colors needed to color a graph G is called chromatic number anDenoted by $\chi(G)$.

Definition 2 Let $P_j(k)$ be a family of all *j*-elemental subsets of $\{1, 2, ..., k\}$. A *j*-fold coloring of graph $G = (V, E)$ is an assignment of *j*-element sets of colors to the vertices of G, *i.e.* $f : V \longrightarrow P_j(k)$, such that for any two adjacent vertices $v, w \in V$ *we have* $f(v) \cap f(w) = \emptyset$. The smallest number of colors *k* needed for a *j*-fold coloring of a graph *G* is called the *j-fold chromatic number* anDenoted by $\chi_j(G)$. The *fractional chromatic number* is defined to be

$$\chi_f(G) := \inf_{j \in \mathbb{N}} \frac{\chi_j(G)}{j} = \lim_{j \to \infty} \frac{\chi_j(G)}{j}$$

Since a 1-fold coloring of a graph is just a classic coloring

$$\chi_f(G) \leq \chi(G).$$

Now, we will present two possible, equivalent notions of graphs on the Euclidean plane we are considering in this work. The first is due to Exoo [1]. The second is introduced in this paper and is more convenient for our work, except Sect. 3.

Definition 3 *An* ε ε *-unit distance graph*, denoted by *G* ε Gε is a graph whose vertices are all points of the plane, in which two points are adjacent if their distance *d* satisfies $1 - \varepsilon \leq d \leq 1 + \varepsilon$, *i.e.*

$$G_\varepsilon = (\mathbb{R}^2, \{\{x, y\} \subset \mathbb{R}^2 \mid 1 - \varepsilon \leq dist(x, y) \leq 1 + \varepsilon\})$$

Definition 4 A graph $G_{[a,b]}$ is a graph whose vertices are all points of the plane $V = \mathbb{R}^2$, in which two points are adjacent if their distance *d* satisfies $a \leq d \leq b$.

$$G_{[a,b]} = (\mathbb{R}^2, \{\{x, y\} \subset \mathbb{R}^2 \mid a \leq dist(x, y) \leq b\}$$

Note that every *G* ε Gε graph can be defined as $G_{[1-\varepsilon, 1+\varepsilon]}$ and every $G_{[a,b]}$ graph can be defined as *G* ε Gε graph with $\varepsilon = \frac{b-a}{b+a}$. Additionally, it is enough to consider graphs $G_{[1,b]}$ since $G_{[1,b]}$ since $G_{[a,b]} \cong G_{[1,b/a]}$.

3 Coloring of $G_{[a,b]}$

The classic Hadwiger–Nelson problem is considered to be very difficult, in particular giving a better general lower bound than 4. In his article Exoo [1] looked for values of ε such that we can determine the chromatic number of G_ε and he succeed in finding some such values. He proved that for $0.134756... < \varepsilon < 0.138998...$ the exact value of $\chi(G_\varepsilon)$ is 7, and for $\varepsilon > 0.008533...$ we get $\chi(G_\varepsilon) \geq 5$.

His work (including computational experiments) suggested that for small enough ε ε the exact value of the chromatic number of G ε Gε is 7.

Conjecture 1

[1] For any $\varepsilon > 0$ *we have* $\chi(G_\varepsilon) = 7.$.

We will give a partial answer to this conjecture. It appears that the pure positivity of ε ε allow us to establish a lower bound of 5, strictly better than 4. We will use a result by Nielsen [8] and for that we need some additional notions.

Given a coloring of the plane *F*, a triangle $T = xyz$ is a *monochromatic limit triangle* if there is a monochromatic set $\{x_1, y_1, z_1, x_2, y_2, z_2, \ldots\}$ such that $x_n \longrightarrow x,\ y_n \longrightarrow y,\ z_n \longrightarrow z$ and each of the triangles $T_n = x_n y_n z_n$ is similar to *T*.

We will say that triangles *xyz* and $x'y'z'$ are $\varepsilon - close$ if $\|x - x'\| < \varepsilon$, $\|y - y'\| < \varepsilon$ and $\|z - z'\| < \varepsilon$.

Theorem 1

[8] Let *F* be a two-coloring of the plane and let *T* be a triangle. Then *F* admits a monochromatic limit triangle congruent to *T*.

Theorem 2

For any $\varepsilon > 0$ we have $\chi(G_\varepsilon) \geq 5.$

Proof

Let $\varepsilon > 0$ and suppose that $\chi(G_\varepsilon) \leq 4.$ Let *F* be a 4-coloring of G_ε and let *F'* be a 2-coloring of the plane such that each point of color 1 or 2 in *F* is white in F' F′ and each point of color 3 or 4 in F is black in F' . Let $T=xyz$ T=xyz be an equilateral triangle with a side length 1.

From Theorem 1 F' F′ admits a monochromatic limit triangle congruent to T(lets say it is white). So from the definition of monochromatic limit triangle there exists n such that the triangle $T_n = x_n y_n z_n$ is $\frac{\varepsilon}{2}-close$ to T.

The side lengths of $T n$ Tn are within the interval $[1-\varepsilon, 1+\varepsilon]$, hence the sides of T_n are edges of G_ε. Since all vertices of $T n$ Tn are white in F' F′ then each of them is colored with 1 or 2 in F. Since there are three vertices and two colors there is a monochromatic edge, which contradicts the assumption that F is a 4-coloring of $G\varepsilon$ □

We believe that there is still room to prove a better bound.

4 Fractional Chromatic Number of $G_{[a,b]}$

The best known way to find an upper bound for the fractional chromatic number of $G_{[1,1]}$ is by looking for a set in the plane with highest possible density but without vertices in distance equal to 1. The densest such set was presented by Hochbeg and O'Donnell [5]. They presented a proof of the upper bound by a limit argument. In the following theorem we generalize their construction for $G_{[1,b]}$. We give a complex description of the method for any b: not only the upper bound, but also an explicit specification of a sequence of fractional colorings "converging" to the upper bound. This sequence give a finite fractional coloring with "quality" as close to the upper bound as needed.

Theorem 3

If $b \geq 1$1 then $\chi_f(G_{[1,b]}) \leq \frac{\sqrt{3}}{3} \cdot \frac{b+\sqrt{1-x^2}}{x}$, where x is the root of $bx = \frac{\pi}{6} -$ arc sin(x). Moreover there exists a sequence of $(\frac{n}{2(b+1)}-1)^2$ -fold colorings with n_2 colors for $n \geq 1$.

Proof

Note: In order to keep this proof shorter we omit some details of calculations. We will define a set S in the plane with high density but without vertices in distance in [1, b] in the following way. Let set A be an intersection of a disk of unit diameter and a hexagon with common center point as in **Figure**. 1. The shape of A changes when we change the size of the hexagon. In the fractional coloring of $G_{[1,b]}$ we choose the set A in such a way that the ratio between the length of the circular arc y and the segment x is equal to b (it will be explained later).

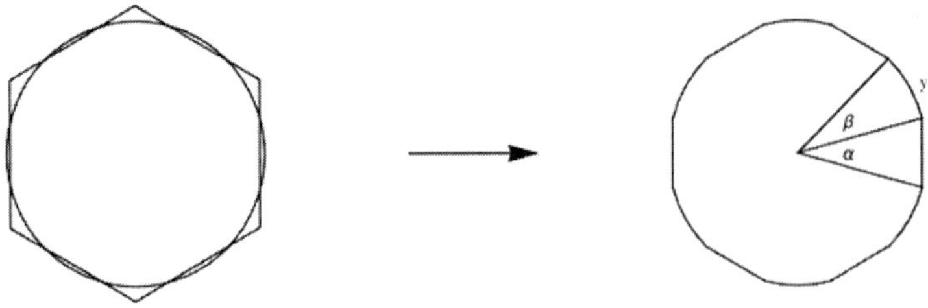

Figure 1. Set *A*

From the construction we get the following equalities $x = \sin(\frac{\alpha}{2}),\ y = \frac{\beta}{2},\ \alpha + \beta = \frac{\pi}{3}$. Therefore we obtain

$$y = \frac{\pi}{6} - \arcsin(x) \tag{1}$$

Then we build *S* by placing copies of *A* on the plane with distance *b* like in **Figure**. 2. Assume that two neighboring components of *S* are centered at (0, 0) and (*s*, 0) which determines the location of all components of *S*. Let *n* be a positive integer. We define a tiling with hexagons of width *s* / *n* in such a way that the left side of some hexagon is a part of the left vertical segment of the border of *A*.

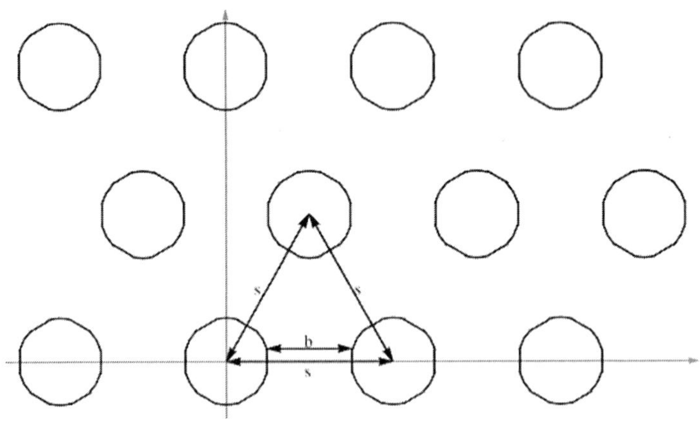

Figure 2. Set *S*

For $0 \leq i, j < n$ let $S_{i,j} = S + (si/n)(1,0) + (sj/n)(1/2, \sqrt{3}/2)$. We define h_n -fold coloring of the plane as follows. We assign color (*i*, *j*) to all points in hexagons fully contained in $S_{i,j}$. Now we will estimate the number h_n of hexagons that are fully contained in one copy of *A*, let say A_0 . Let A_{0n} be a rescaled copy o *A* with the same center as A_0 anDiameter $(1 - 2\frac{s}{n})$. The area of A_{0n} is

A_n^0 is $p_n = \frac{1}{4}(1-2\frac{s}{n})^2(\pi - 6\arcsin(x) + 6x\sqrt{1-x^2})$, and the area of the hexagon is $(s/n)^2(\sqrt{3}/2)$. Observe that if a hexagon of the tiling has non-empty intersection with A_n^0 An0 then it is fully contained in A^0 A0. Therefore we can bound h_n hn from below:

$$h_n \geq \frac{p_n}{(s/n)^2(\sqrt{3}/2)} = \frac{\frac{1}{4}(1-2\frac{s}{n})^2(\pi - 6\arcsin(x) + 6x\sqrt{1-x^2})}{(s/n)^2(\sqrt{3}/2)} =$$

$$= \sqrt{3}(\frac{n}{b+\sqrt{1-x^2}} - 2)^2(bx + x\sqrt{1-x^2}).$$

On the other hand we conclude that every hexagon is contained in h_n of $S_{i,j}$ (since we can get all the hexagons by shifting one of them by $(si/n)(1,0) + (sj/n)(1/2, \sqrt{3}/2))$, hence every point in a hexagon has h_n hn of n^2 n2 colors assigned.

With n tending to infinity the set of points colored 1 tends to S and our sequence of h_n -fold colorings gives us an upper bound for the fractional chromatic number of $G_{[1,b]}$:

$$\chi_f(G_{[1,b]}) \leq \lim_{n\to\infty} \frac{n^2}{\frac{1}{2\sqrt{3}}\left(\frac{n}{b+\sqrt{1-x^2}} - 2\right)^2\left(\pi - 6\arcsin(x) + 6x\sqrt{1-x^2}\right)} =$$

$$= \frac{2\sqrt{3}}{\pi - 6\arcsin(x) + 6x\sqrt{1-x^2}} \cdot \lim_{n\to\infty} \frac{n^2}{\left(\frac{n}{b+\sqrt{1-x^2}} - 2\right)^2} =$$

$$= \frac{2\sqrt{3}(b+\sqrt{1-x^2})^2}{\pi - 6\arcsin(x) + 6x\sqrt{1-x^2}} =$$

$$= \frac{2\sqrt{3}(b+\sqrt{1-x^2})^2}{\pi - 6\arcsin(x) + 3\sin(2\arcsin(x))}$$

Now let us explain why did we choose the ratio between the length of the circular arc—y and the segment—x to be b. To find minimal value we take the derivative of the upper bound for $\chi_f(G_{[1,b]})$ presented above and check when it equals 0.

$$\frac{4\sqrt{3}x\left(b+\sqrt{1-x^2}\right)\left(6bx + 6\arcsin(x) - \pi\right)}{\sqrt{1-x^2}\left(-6\arcsin(x) + 3\sin\left(2\arcsin(x)\right) + \pi\right)^2} = 0$$

$$\Longleftrightarrow 6bx + 6\arcsin(x) - \pi = 0$$

Applying equality 1 we obtain

$6bx - 6y = 0$ and hence

$y = bx$

To complete the proof we transform the formula $\frac{2\sqrt{3}(b+\sqrt{1-x^2})^2}{\pi - 6\arcsin(x) + 3\sin(2\arcsin(x))}$) using the previous equality and by doing so we get $\frac{\sqrt{3}}{3} \cdot \frac{b+\sqrt{1-x^2}}{x}$ as the upper bound for the fractional chromatic number of graph $G_{[1,b]}$. □ □

Figure 3 and Table 1 present the upper bound for the fractional chromatic number of *G* [1,*b*] G[1,b] for different values *b*.

$b =$	1	1,5	2	3	4
$\chi_f(G_{[1,b]}) \leq$	4,36	6,86	9,9	17,62	27,55

Table 1: Applications of Theorem3

5 *j*-Fold Coloring of the Plane

Note that the upper bounds for the fractional chromatic numbers of $G_{[1,b]}$ are established by presenting an infinite sequence of *j*-fold colorings. These colorings give results close to the upper bound only for very big *j*. However, as it was mentioned in the introduction, for the practical reasons it is often the case that only *j*-fold coloring with small *j* are valuable to consider. Furthermore, this consideration leads also to a purely mathematical question: How fast, in terms of *j*, can we get close to the infinite limit - the upper bound for the fractional chromatic number? In this section, we give some insight into this matter.

5.1 Methods for $G_{[1,1]}$

In this subsection, we concentrate on $G_{[1,1]}$.

Theorem 4

There are twofold coloring of $G_{[1,1]}$ with 12 colors and threefold coloring of $G_{[1,1]}$ with 16 colors.

Proof

For our colorings we are going to use classic covering of the plane with hexagons with side length 1 / 2. Obviously in a hexagon of side length 1 / 2 there are pairs of point in a distance 1, so when we say we color a hexagon we mean the interior of the hexagon plus its right border and its three vertices: the upper one and the two on the right (see **Figure**. 4).

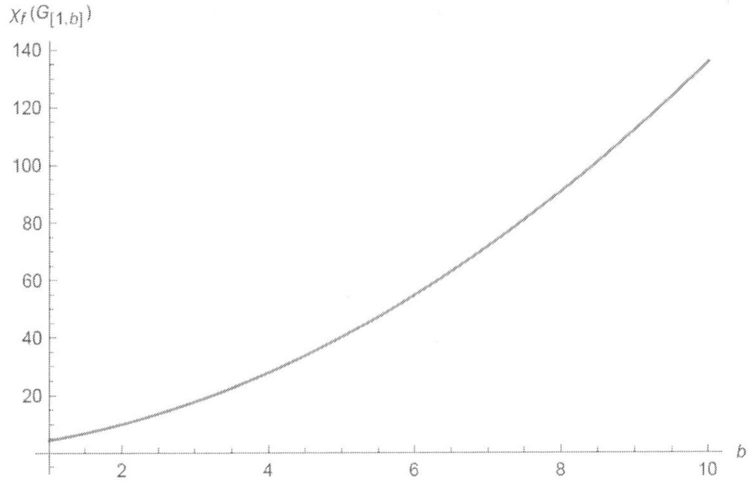

Figure 3. The upper bound for $\chi_f(G_{[1,b]})$ from Theorem 3

Firstly we show our twofold coloring of $G_{[1,1]}$ with 12 colors. We are going to use the hexagon grid twice with the set of colors $\{1,2,\ldots,12\}$. We create the first layer by simply giving each row of hexagons three numbers and use it periodically so that if one row has colors from the set $\{1,2,3\}$ the next gets colors $\{4,5,6\}$ and so on. Having this coloring as our model we get the second layer by moving the colored grid by a vector $[3\sqrt{3}/4, -3/2].$. See **Figure**. 5 with the first layer of colors and marked placement of color 1 in the layers. The distances between two hexagons of the same color are: vertically 2, horizontally $\sqrt{3}$ anDiagonally $\frac{5\sqrt{3}}{8} \approx 1.08,$, so they are all bigger than 1.

To create a threefold coloring of $G_{[1,1]}$ with 16 colors we again start with the classic covering of the plane with hexagons with side length 1/2. We are going to use the hexagon grid 3 times with the set of colors $\{1,2,3,\ldots,16\}$ .We create the first layer by simply giving each row of hexagons four numbers and use it periodically so that if one row has colors from the set $\{1,2,3,4\}$ the next gets colors $\{5,6,7,8\}$ and so on. Having this coloring as our model we get second and third layers by moving the colored grid by a vector $[\sqrt{3}, -1].$. See **Figure**. 6 with the first layer of colors and marked placement of color 1 in other layers. The distances between two hexagons of the same color are at least one (exactly one in case of two hexagons from different layers). □ □

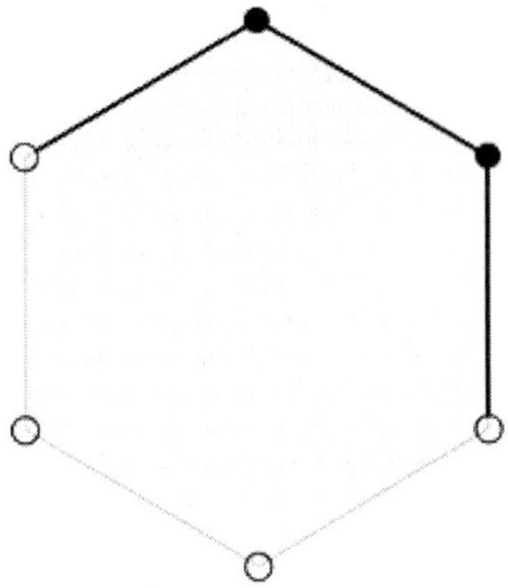

Figure 4. Borders of a hexagon

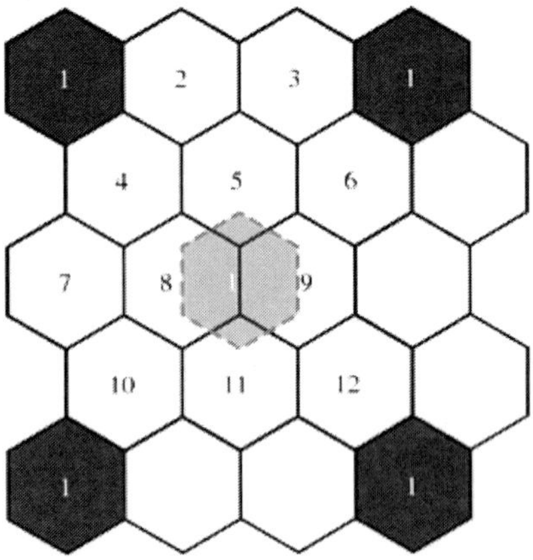

Figure 5. Twofold coloring of $G_{[1,1]}$ with 12 colors

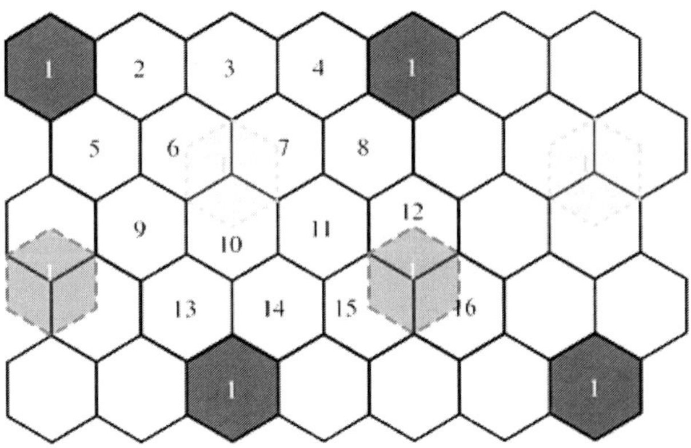

Figure 6. Threefold coloring of $G_{[1,1]}$ with 16 colors

Theorem 5

There is a sevenfold coloring of $G_{[1,1]}$ with 37 colors i.e. $\frac{37}{7} \approx 5.285.$

Proof

To create sevenfold coloring of $G_{[1,1]}$ we start with a hexagon grid with hexagon side's length $s = \frac{1}{2\sqrt{7}}.$ We color with the first color some hexagons in the pattern presented as shadowed on **Figure**. 7. Then we shift our pattern by a vector $[\frac{\sqrt{3}}{2\sqrt{7}}, 0]$ and color it with second color (bold contour in **Figure**. 7). Repeating this action 37 times we have 7 colors for each hexagon and there are 37 colors total. In our base coloring **Figure**ure made by 7 hexagons the largest distance between two points is 1 so choosing half of the border to be colored and the other not is enough to make sure there are no two point in distance 1 in one **Figure** The distance between two such **Figure**ures is

$$\sqrt{(3\sqrt{3}s)^2 + (2s)^2} = s\sqrt{31} = \frac{\sqrt{31}}{2\sqrt{7}} \approx 1.05$$

□ □

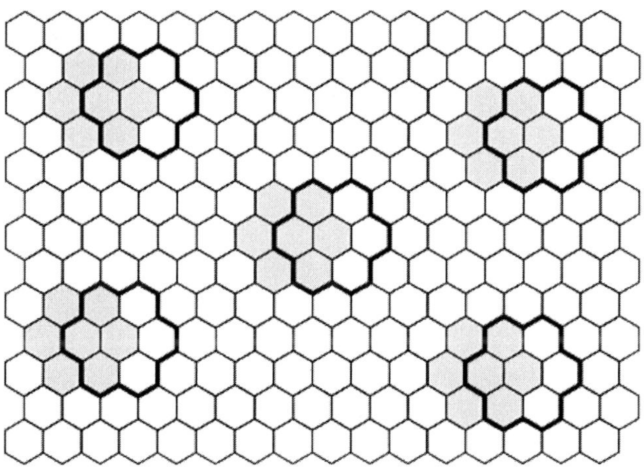

Figure 7 Sevenfold coloring of $G_{[1,1]}$ with 37 colors

5.2 Methods for $G_{[a,b]}$

In this section we give two general methods for building *j*-fold colorings (for small *j*) for graphs $G_{[1,b]}$.

Theorem 6

There exists a *nm*-fold coloring with $\lceil(\frac{2b}{\sqrt{3}}+1)\cdot n\rceil\cdot\lceil(\frac{2b}{\sqrt{3}}+1)\cdot m\rceil$ $G_{[1,b]}$ colours of the graph i.e. $\frac{\chi_{nm}(G_{[1,b]})}{nm} \leq \frac{\lceil(2b/\sqrt{3}+1)\cdot n\rceil\cdot\lceil(2b/\sqrt{3}+1)\cdot m\rceil}{nm}$.

Proof

In the proof we are going to create *n·m* n·m colored hexagon grids. A color of a hexagon will be a pair of numbers, first of which will be related to the row the hexagon is in and the second corresponds to the column.

Let W_1^1 be a hexagon grid with hexagons with side length 1/2. Let *H* be one of the hexagons from W_1^1. For $2 \leq j \leq n$ let W_1^j be a hexagon grid created by moving uncolored W_1^1 by a vector $(j-1)/n\ [\sqrt{3}/2, 0]$. Now, lets say *H* is colored (1, 1) and let's find first hexagon *H'* H′ in the same row from any W_1^j that can also be colored (1, 1) without creating a monochromatic edge in $G_{[1,b]}$ i.e at distance greater or equal to *b* (see **Figure**. 8).

Counting the hexagons from *H* to *H'* we find that there are exactly $\lceil(b+\frac{\sqrt{3}}{2})\frac{2n}{\sqrt{3}}\rceil = \lceil(\frac{2b}{\sqrt{3}}+1)\cdot n\rceil$ of them, since the distance between the centers of *H* and *H'* has to be at least $b+\frac{\sqrt{3}}{2}$. Then we shift each of W_1^j by vectors $(i-1)/m\ [\sqrt{3}/4, -3/4]$ for $2 \leq i \leq m$ getting grids W_1^j. Now remembering that *H* is colored (1, 1) we find the first hexagon *H"* from W_1^j to have the same color (see **Figure**. 9). The number of hexagons between the two is $\lceil(2b/\sqrt{3}+1)\cdot m\rceil$.

Now we have three hexagons *H*, *H'*, and *H"* that can be colored with the same color, since the distance between *H'* and *H"* would be the smallest when *n*=*m* n=m and then the centers of the three hexagons create an equilateral triangle with side length at least $b+\frac{\sqrt{3}}{2}$ (because *H* and *H'* are in proper distance).

The final coloring is created by giving each hexagon a pair of numbers, so that *H* gets color (1, 1), the next hexagon in the same row gets (1, 2) and so on until *H'* H′ gets (1, 1) again and the cycle repeats, and in the row below we get colors (2,1), (2,2), (2,3)… (2,1), (2,2), (2,3)… and so on until we have row with *H"* H″ in which we use (1,1), (1,2), (1,3)… (1,1), (1,2), (1,3)… again. Since we have *nm* hexagon grids every point gets *nm* colors. Finally we get *nm*-fold coloring of $G_{[1,b]}$ with $\lceil(\frac{2b}{\sqrt{3}}+1)\cdot n\rceil\cdot\lceil(\frac{2b}{\sqrt{3}}+1)\cdot m\rceil$ colors (**Figure**. 10). □ □

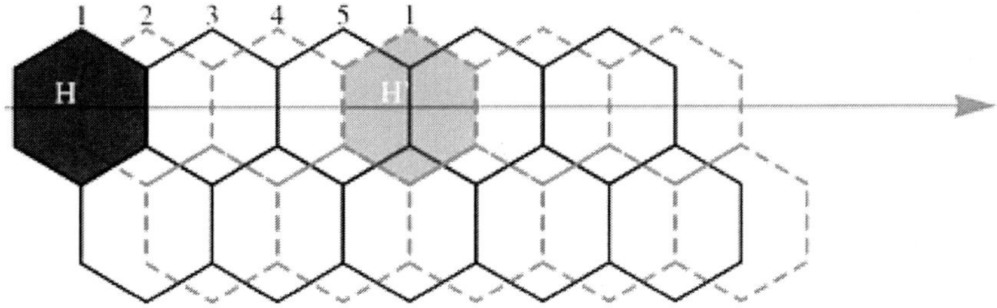

Figure 8 Twofolds in fourfold coloring $G_{[1,1]}$ with 25 colors (different *shades/colors* represent color (1, 1) from different layers)

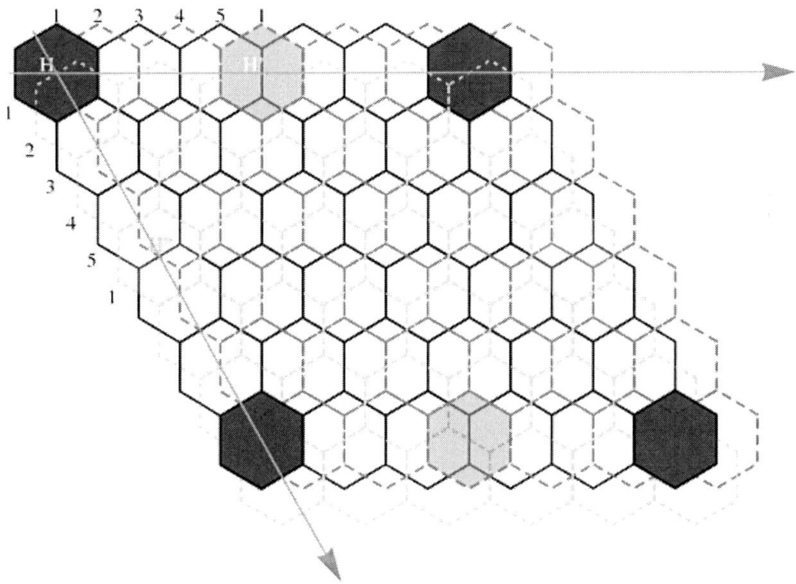

Figure 9 Threefolds in fourfold coloring $G_{[1,1]}$ with 25 colors (different *shades/colors* represent color (1, 1) from different layers)

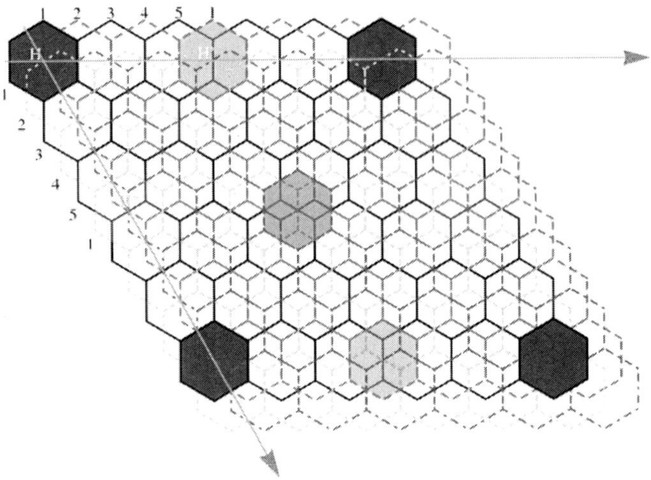

Figure 10. Fourfold coloring $G_{[1,1]}$ with 25 colors (different *shades/colors* represent color (1, 1) from different layers)

The following result can be seen as a combination of twofold coloring approach from Theorem 4 and the main method from Theorem 6.

Theorem 7 There exists a *2nm*-fold coloring with $2\lceil (b+1)2n \rceil \lceil (b+1)\frac{2m}{3} \rceil$ colors of the graph $G_{[1,b]}$ $i.e.$ $\frac{\chi_{2nm}(G_{[1,b]})}{2nm} \leq \frac{2\lceil \sqrt{3}(b+1)\frac{2n}{3} \rceil \lceil (b+1)\frac{2m}{3} \rceil}{2nm}$.

Proof

Note that the proof uses similar methods to the previous one. We are going to create 2·*n*·*m* 2·n·m colored hexagon grids. A color of a hexagon will be a pair of numbers, first of which will be related to the row the hexagon is in and the second corresponds to the column.

Let W_1^1 be a hexagon grid with hexagons with side length 1/2. Let *H* be one of the hexagons from W_1^1. For $2 \leq i \leq m$ let W_i^1 be a hexagon grid created by moving uncolored W_1^1 by a vector $(i-1)/m\ [0,-3/2]$. . Now lets say *H* is colored (1, 1) and let's find first hexagon *H'* H' in the same column from any $W^1{}_i$ that can also be colored (1, 1) without creating a monochromatic edge in $G_{[1,b]}$ i.e at distance greater or equal to *b*, so the distance between the centers of *H* and *H'* needs to be at least *b*+1 (see **Figure.** 11 for fourfold coloring with 4·6=24 4·6=24 colors of $G_{[1,1]}$ with $n = 1$ and $m = 2$) Since the distance between the centers of H and H' is at least b+1 but we chose H' to be as close as possible to H, then there are $\lceil (b+1)\frac{2m}{3} \rceil$ rows between H and H'. Now for $2 \leq j \leq m$ and $2 \leq j \leq n$ let W_i^j be a hexagon grid created by shifting $W1i$ Wi1 by a vector $2 \leq j \leq m$ and $2 \leq j \leq n$. Let H'' be the first hexagon in the same row as H in any of W_i^j such that the distance between the centers of the two is at least $\sqrt{3}(b+1)$. The number of hexagons between H and H'' is $\lceil \sqrt{3}(b+1)\frac{2n}{\sqrt{3}} \rceil = \lceil (b+1)2n \rceil \geq b + \frac{\sqrt{3}}{2}$.

The distance between the centers of H' and H'' is at least:

$$\sqrt{\left(\lceil (b+1)\frac{2m}{3} \rceil \frac{3}{2m} \right)^2 \cdot \left(\lceil \sqrt{3}(b+1)\frac{2n}{\sqrt{3}} \rceil \frac{\sqrt{3}}{2n} \right)^2} \geq$$

$$\geq \sqrt{(b+1)^2 + (\sqrt{3}(b+1))^2} = \sqrt{4(b+1)^2} = 2b+2.$$

Since the distance is at least 2*b*+2 they can be colored the same color and there is enough space between them to put another hexagon in the same color between them. So we create new hexagon grids V_i^j by shifting W_i^j by a vector $[\lceil \sqrt{3}(b+1)\frac{2n}{\sqrt{3}} \rceil \cdot \frac{\sqrt{3}}{4n}, \lceil (b+1)\frac{2m}{3} \rceil \cdot \frac{3}{4m}]$. The final coloring is created by giving each hexagon a pair of numbers, so that H gets color (1, 1) the next hexagon in the same row from any W_1^j gets (1, 2) and so on until H'' gets (1, 1) again and the cycle repeats. The next row from W_1^j has the first coordinate in all colors equal to 2. The set of numbers for second coordinate is $\{1, 2, ..., \lceil \sqrt{3}(b+1)\frac{2n}{\sqrt{3}} \rceil\}$, and for the first coordinate $2\lceil (b+1)\frac{2m}{3} \rceil$ (since there are 2 rows of hexagons between 2 hexagons from the same grid in the same column). So we use $2\lceil (b+1)\frac{2m}{3} \rceil \cdot \lceil \sqrt{3}(b+1)\frac{2n}{\sqrt{3}} \rceil$ colors using 2*nm* hexagon grids. □ □

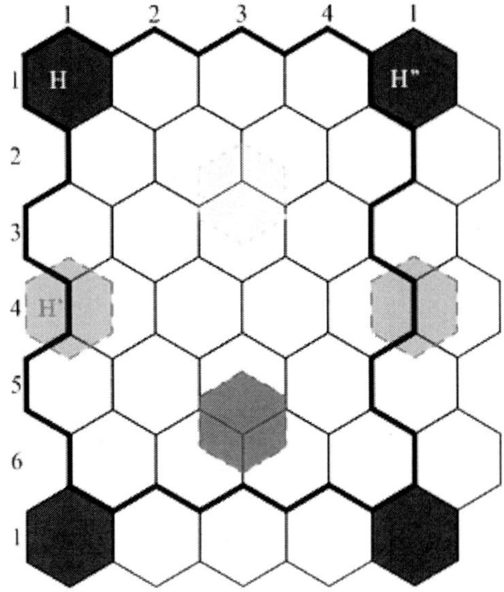

Figure 11. Fourfold coloring $G_{[1,1]}$ with 24 colors (different *shades/colors* represent color (1, 1) from different layers)

5.3 SUMMARY

Theorems 6 and 7 give different results and we cannot say one is stronger than another. Table 2 presents comparison of the two for *j*-fold coloring of $G\,[1,1]$ G[1,1] with small values of *j* (*k* is the numbers of colors used). We bold best results for a fixed *j*.

Table 2. Applications of Theorems 6 and 7 for $G_{[1,1]}$ with small *j*

method:	$j =$	2	4	6	8	10	12
j=nm	$k =$	**15**	25	35	**45**	**55**	**63**
	$k/j \approx$	7,5	6,25	5,83	5,63	5,5	5,25
j=2nm	$k =$	16	**24**	**32**	48	56	64
	$k/j \approx$	8	6	5.33	6	5,6	5,33

Our best results in *j*-fold coloring of $G\,0$ with small values of *j* using *k* colors are summarized in Table 3. They follow from Theorems 4, 5, 6, 7.

Table 3. Selected results in *j*-fold coloring of $G_{[1,1]}$ with small *j*

k =	7	12	16	24	32	32	37
j =	1	2	3	4	5	6	7
k/j ≈	7	6	5.33	6	6.6	5.33	5.26

For practical applications it is useful to consider *j*-fold coloring of graph $G_{[1,2]}$, especially with small *j*. Table 4 presents our results in coloring $G_{[1,2]}$ using method from Theorem 6. It appears that in this case the method from Theorem 7 does not give good results.

Table 4. Applications of Theorem 6 for $G_{[1,2]}$ with small *j*

k =	12	70	100	930	960
j =	1	6	9	84	87
k/j ≈	12	11,67	11,11	11,07	11,03

ACKNOWLEDGMENTS

We thank Professor Zbigniew Lonc, Zbigniew Walczak and Professor Jacek Wojciechowski for introducing us to the problem.

REFERENCES

1. Exoo, G.: ε ε -Unit distance graphs. Discrete Comput. Geom. **33**, 117–123 (2005)

2. Falconer, K.J.: The realization of distances in measurable subsets covering R *n* Rn. J. Comb. Theory, Ser. A **31**, 187–189 (1981)

3. Fisher, D., Ullman, D.: The fractional chromatic number of the plane. Geombinatorics **2**(1), 8–12 (1992)

4. Hadwiger, H.: Ungelöste Probleme. Elem. Math. **16**, 103–104 (1961)

5. Hochberg, R., O'Donnell, P.: A large independent set in the unit distance graph. Geombinatorics **3**(4), 83–84 (1993)

6. Ivanov, L.L.: On the chromatic numbers of R 2 R2 and R 3 R3 with intervals of forbidden distances. Electron. Notes Discrete Math. **29**, 159–162 (2007)

7. Moser, L., Moser, W.: Solution for problem 10. Can. Bull. Math. **4**, 187–189 (1961)

8. Nielsen, M.J.: Approximating monochromatic triangles in a two-colored plane. Acta Math. Hung. **74**(4), 279–286 (1997)

9. Scheinerman, E.R., Ullman, D.H.: Fractional Graph Theory. Wiley, New York (2008)

10. Soifer, A.: The Mathematical Coloring Book. Springer, New York (2008)

11. Walczak, Z., Wojciechowski, J.M.: Transmission scheduling in packet radio networks using graph coloring algorithm, wireless and mobile communications. In: ICWMC '06 (2006)

12. Woodall, D.R.: Distances realized by sets covering the plane. J. Comb. Theory, Ser. A **74**(4), 279–286 (1997)

CHAPTER 8

Unions of Fat Convex Polytopes Have Short Skeletons

Boris Aronov, Mark de Berg

Department of Computer Science and Engineering, Polytechnic Institute of NYU

Department of Computing Science, TU Eindhoven

ABSTRACT

The skeleton of a polyhedral set is the union of its edges and vertices. Let P P be a set of fat, convex polytopes in three dimensions with n vertices in total, and let f_{max} be the maximum complexity of any face of a polytope in P P. We prove that the total length of the skeleton of the union of the polytopes in P P is at most $O(\alpha(n) \cdot \log^* n \cdot \log f_{max})$ times the sum of the skeleton lengths of the individual polytopes.

Keywords: Fat polytopes Convex polytopes Skeleton of union Combinatorial complexity

1 INTRODUCTION

Background and Motivation

Many geometric algorithms require computing the union of a collection of objects in the plane or some higher-dimensional space. For this reason there are many papers studying the *combinatorial complexity* of the union—that is, the total number of vertices, edges, and higher-dimensional faces of the union; see the recent survey by Agarwal et al. [1] and references therein.

In the plane, the worst-case combinatorial complexity of the union of n constant-complexity objects (triangles, rectangles, etc.) is easily seen to be $\Theta(n^2)$. However, if the objects are *fat*—see below for a formal definition—then the union complexity is near linear [1, 8, 10, 11, 13]. More precisely, the current best bound on the union complexity of n β-fat triangles is $O(n\alpha(n)\log^* n)$ [13], with the constant of proportionality depending on the fatness parameter β; here $\alpha(n)$ is the inverse Ackermann function and $\log^* n$ denotes the iterated logarithm. (Thus, $\log^* n=0$ for $n\leq 1$ and $\log^* n=1+\log^*(\log n)$, with logarithms taken base 2.) For fat planar objects with curved boundaries a near-linear bound is known as well [1, 2, 5]. In $\mathbb{R}^3$, the union complexity of n constant-complexity objects is $\Theta(n^3)$. The general belief is that, as in the planar case, the union complexity of n fat objects in $\mathbb{R}^3$ is roughly one order of magnitude smaller than that of general objects, namely near quadratic. Until recently this result was known only for rather restricted classes of objects such as similarly sized cubes (of arbitrary orientation) and fat wedges [12]. Then Ezra and Sharir [7] obtained a breakthrough in this area by showing that the union complexity of n fat tetrahedra, not necessarily of similar size, is $O(n^{2+\varepsilon})$; this implies an analogous result for cubes. However, for fat polytopes near-quadratic bounds are still not known—the main problem being that fat polytopes can have non-fat faces.

The combinatorial complexity of the union of objects in $\mathbb{R}^3$ is proportional to the number of vertices and edges of the union—in other words, the complexity of the skeleton of the union. In order to further increase our understanding of the union of fat objects in $\mathbb{R}^3$, we study a different aspect, namely, the total *length* of the skeleton, rather than the number of edges and vertices it is composed of. As with combinatorial complexity, the main question is whether the worst-case skeleton length of the union of fat objects is smaller than that of general objects. Consider, for example, a collection of thin "plates" in $\mathbb{R}^3$, where each plate is a slightly inflated unit square. If one arranges these plates in a grid-like manner, the skeleton length is $\Theta(n^2)$, which is significantly larger than the total length $\Theta(n)$ of the skeletons of the plates. Now suppose that instead of thin plates we take a collection of n unit cubes. Can these also be arranged in such a way that the skeleton length of their union is $\Theta(n^2)$? If the cubes are *axis-parallel* then this cannot be done—this follows easily from the fact that the union complexity of n axis-parallel unit cubes is $O(n)$ [3]. But what about arbitrarily oriented unit cubes? And what can be said about other types of fat objects?

Our Result

We prove that the answer to the question posed above is negative: the maximum length of the skeleton of the union of n (arbitrarily oriented) unit cubes is only $O(n\alpha(n)\log^* n)$. Our main result is, in fact, much more general than this, as it applies to arbitrary fat convex polytopes. Before we state our result more precisely, we need to introduce some terminology and notation. From now on, whenever we will speak of polytopes, we always mean convex polytopes.

Let $\mathrm{vol}(o)$ denote the volume of an object o. An object o in $\mathbb{R}^d$ is called *β-fat*, for some $\beta>0$, if for any ball b whose center lies in o and that does not fully contain o in its interior, we have $\mathrm{vol}(b\cap o)\geq\beta\cdot\mathrm{vol}(b)$ [15, 16]. The *fatness* of an object o is the largest value β such that o is β-fat. For objects that are convex and bounded—the case we consider in this paper—it can be shown that the fatness is determined by a ball whose radius is the object's diameter [15, 16]; thus the fatness of a convex bounded object o is equal to $\mathrm{vol}(o)/(\omega_d \, \mathrm{diam}(o)^d)$, where $\mathrm{vol}(o)$ and $\mathrm{diam}(o)$ denote the volume anDiameter of o, respectively, and ω $_D$enotes the volume of the d-dimensional unit ball. From this it follows that the maximum value of the fatness is attained when o is a ball, in virtue of the isodiametric inequality [4]. This maximum value is $1/2^d$, that is, 1/8 for d=3.

The *skeleton* of a polyhedral object S, which we denote by $\mathrm{skel}(S)$, is the union of all of its edges and vertices. Slightly abusing the notation, we sometimes also refer to the *set* of the edges and vertices of S as the skeleton of S. We use $\|\mathrm{skel}(S)\|$ to denote the *total length* of the skeleton of S, i.e., the sum of the lengths of its constituent edges. We also use the notation $\|\cdot\|$ to denote the length of an edge or an interval; the notation $|\cdot|$ will be used to denote the cardinality of a set. Our main result can now be stated as follows.

Theorem 1

Let $0<\beta\leq 1/8$ be a fixed constant, and let P *be a set of convex β-fat polytopes in $\mathbb{R}^3$ with n vertices in total. Let $f_{\max}$ be the maximum complexity of any face of a polytope in* P *. Then*

$$\left\|\mathrm{skel}\left(\bigcup \mathcal{P}\right)\right\| \leqslant K_\beta \cdot \sum_{P \in \mathcal{P}} \|\mathrm{skel}(P)\| \cdot \alpha(n) \log^* n \cdot \log f_{\max},$$

where K_β is a constant whose value depends only on the fatness constant β.

Furthermore, there exists a set P P *of n polytopes, each of fatness $\Theta(1)$, constant complexity, anDiameter 1, with* $\left\|\mathrm{skel}\left(\bigcup \mathcal{P}\right)\right\| = \Omega(n\alpha(n))$; *here* $\alpha(\cdot)$; *here $\alpha(\cdot)$ is the inverse Ackermann function.*

Remark

The $\alpha(n)\log^* n$ factor in the bound is due to the fact that the best published bound on the union complexity of n fat triangles in the plane is $O(n\alpha(n)\log^* n)$ [13]. Any improvement in this union bound will immediately imply an improvement in the bound of Theorem 1. In a recent manuscript [2] a new bound of $O(n\log^* n)$ has been proved for the union complexity of fat triangles. This implies that the $\alpha(n)\log^* n$ factor in Theorem 1 can be reduced to $\log^* n$.

2 THE PROOF

We first consider the special case where P P consists of n convex fat polytopes of constant complexity each, and then extend the result to convex fat polytopes whose total complexity is n. We assume without loss of generality that the set of polytopes is in general position; in particular, no polytope vertex lies on the boundary of another polytope and no two edges of different polytopes intersect.

2.1 Constant-Complexity Polytopes

Let $\mathcal{P} = \{P_1, \ldots, P_n\}$ be a set of n convex β-fat polytopes, each with at most C faces. The main idea of the proof is as follows. Suppose we sweep a plane through the collection of polytopes. A union vertex of the cross section then corresponds to an edge of the skeleton skel($\bigcup_i P_i$) intersecting the sweep plane. Thus we can bound the skeleton length by integrating the number of union vertices in the cross section over the sweep, where each vertex is weighted suitably, depending on the angle its corresponding skeleton edge makes with the sweep plane.

To get a good bound using this approach, the complexity of the cross sections of the union needs to be small. This would be the case if the polytope cross sections were fat, since the union of fat objects in the plane has near-linear complexity. However, the cross section of a fat polytope P with a plane need not always be fat. For example, if the sweep plane is parallel to an edge e of P, and e is the first (or last) encountered feature of P, then the cross section will be arbitrarily skinny just after (or just before) hitting e. We start by showing that this is the only way a cross section of a fat polytope can be non-fat. Namely, if the angle of the sweep direction with all polytope edges is bounded away from $\pi/2$—so that the angle of the sweep plane with each edge is bounded away from zero—then the cross sections are fat. We then use this fact to bound the total skeleton length of the union by performing a number of sweeps, each involving only polytopes whose cross sections are fat, and summing the contributions of the sweeps to the skeleton length.

We use $\measuredangle(h,e)$ to denote the measure of the angle between a plane h and a segment e, and $\measuredangle(\phi,e)$ to denote the measure of the angle between a direction ϕ and a segment e. Here segments are not oriented; opposite orientations are considered the same direction. Note that the values $\measuredangle(h,e)$ and $\measuredangle(\phi,e)$ are always in the range $[0,\pi/2]$.

Definition

A direction ϕ is *β-fattening* for a convex polytope P if every intersection of P with a plane orthogonal to ϕ and meeting the interior of P is a β-fat polygon.

Lemma 2

Let P be a β-fat convex polytope in $\mathbb{R}^3$, let $\varepsilon>0$ be a constant, and let ϕ be a direction with $\measuredangle(\phi, e) \leqslant \pi/2 - \varepsilon$ for every edge e of P. Then ϕ is β'-fattening for P, for $\beta' := \frac{\beta \tan^2 \varepsilon}{3}$.

Proof

Let h be a plane orthogonal to ϕ and meeting the interior of P. Put $d{:=}\text{diam}(P\cap h)$ and $A{:=}\text{area}(h\cap P)$. Since the fatness of a convex object is determined by a ball whose radius is the object's diameter [15, 16] it is sufficient to show that $A\geq\beta'\cdot\pi d^2$.

Assume without loss of generality that h is horizontal. Consider the area of intersection with P as we sweep a horizontal plane $h(z)$ from $z=-\infty$ to $z=+\infty$. Then $\text{area}(h(z)\cap P)^{1/2}$ is a concave function of z, by Brunn's theorem [4]. Thus either $\text{area}(h(z)\cap P)\leq A$ for any plane $h(z)$ below h, or $\text{area}(h(z)\cap P)\leq A$ for any plane $h(z)$ above h; we assume without loss of generality that the latter is the case. Let v be the vertex of p that is above h and furthest away from h—see **Figure**. 1. Since P does not have edges parallel to h, the vertex v is unique.

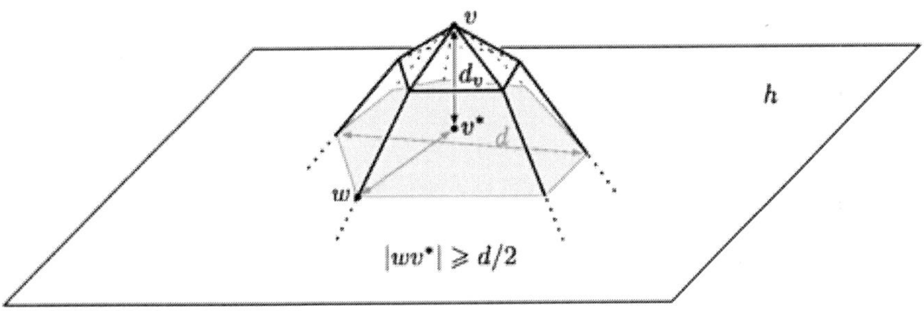

Figure. 1. Sweeping the polytope with a plane h

Let d_v be the distance between h and v, let P^+ be the part of P lying above h, and let B_v be the ball with center v and radius d_v. Since P is β-fat, we have

$$\beta \cdot \frac{4}{3}\pi d_v^3 \leqslant \text{vol}(B_v \cap P) = \text{vol}(B_v \cap P^+) \leqslant \text{vol}(P^+) \leqslant A \cdot d_v.$$

It follows that

$$A \geqslant \beta \cdot \frac{4}{3}\pi d_v^2. \qquad (1)$$

Define v^* to be the vertical projection of v onto the plane h. Then $P\cap h$ has a vertex w with $\|wv^*\|\geq d/2$. We claim that this implies that P has an edge e such that $\tan\measuredangle(h,e)\leq 2d_v/d$. To see this, walk from w to v along the edges of P in such a way that the height along the path increases monotonically. This is always possible since P is convex anDoes not have edges parallel to h. The length of the projection of this path onto h is at least $\|wv^*\|$ and the height increase along the path is d_v. Hence, there is an edge e with slope at most $d_v/\|wv^*\|\leq 2d_v/d$, as claimed, and we have

2d v /d≥tan∡(h,e)≥tanε. 2dv/d≥tan∡(h,e)≥tanε.

Combining with (1), we obtain

$$A \geqslant \beta \cdot \frac{4}{3}\pi d_v^2 \geqslant \beta \cdot \frac{4}{3}\pi (\frac{1}{2}d \tan \varepsilon)^2 = \frac{\beta \tan^2 \varepsilon}{3} \cdot \pi d^2,$$

which proves the lemma. □

The following result was proved by Pettie [13] for fat triangles; by the arguments of De Berg and Gray [6], it can be extended to fat polygons.

Fact 3

Given a collection of convex, β-fat polygons in the plane with a total of n edges, the number of vertices of their union is at most $f_\beta(n) := O(n\alpha(n)\log^* n + (n/\beta)\log^2(1/\beta)))$.

We consider *β* to be a fixed constant, and will not analyze the exact dependency in our bounds on *β*. Hence, we simply use *O*(*nα*(*n*)log* *n*) as an upper bound on the union complexity.

The main ingredient for the proof of the version of Theorem 1 for constant-complexity polytopes is the following lemma.

Lemma 4

Let 0<*β*≤1/8 *and* C≥4 *be constants. There exists a set Φ of O*(C^4) *directions with the following property: for any finite set* P P *of convex β-fat polytopes in* $\mathbb{R}^3$ *in general position, each with at most C faces, there exists, for every edge e of* skel(UP) skel(UP), *a direction* ϕ_e∈*Φ such that*

1. (P1) $\measuredangle(\phi, e) \leqslant \pi/3,$ *and*

2. (P2)*ϕ is β'-fattening for the at most two polytopes of* P P *whose boundary contains e, where* $\beta' := \frac{\beta}{3(48C)^2}.$

Proof

We will concentrate on proving the existence of the set *Φ* of directions with the required properties and not focus on minimizing its size as a function of the various parameters. We will identify the set of all orientations with the unit sphere S S; technically, opposite points on S S represent the same direction, but this will not affect the argument below.

Define

$$\rho := \frac{\pi}{16\binom{6C}{2} + 16}$$

and take Φ to be a set of points on S S such that any spherical cap of radius ρ contains at least one point from Φ. It is well known that there exists such a set of size $O(1/\rho^2)=O(C^4)$. Next we prove that Φ has the desired properties.

Consider an edge e of skel(UP) skel⁡(UP). Assume e is defined by two polytopes P_1 and P_2, that is, e is contained in the intersection of faces f_1 of P_1 and f_2 of P_2; the argument for the case where e is contained in an edge of a single polytope is easier. Let $\xi \in S$ ξ∈S be the direction of e. Any direction ϕ in the spherical cap C_ξ of area π and radius $\pi/3$ centered at ξ satisfies property (P1).

Now consider property (P2). Let ε be such that $0<\varepsilon<\pi/2$ and $\sin\varepsilon=148C$ sin⁡ε=148C. Since $\tan^2\varepsilon > \sin^2\varepsilon = \frac{1}{(48C)^2}$, any direction ϕ that makes an angle at most $\pi/2-\varepsilon$ with every edge s of some β-fat polytope P is β'-fattening for P, by Lemma 2. This means a direction $\phi \in S$ ϕ∈S is β'-fattening for P if it stays out of a band of half-width ε centered at the great circle of directions perpendicular to that of s, for every edge s of P. Since P_1 and P_2 have at most C faces, each has fewer than $3C$ edges. Hence there are fewer than $6C$ forbidden bands. The area of each band is $4\pi\sin\varepsilon$. The complement of the bands inside C_ξ consists of several of spherically convex regions—see **Figure**. 2. The number of such regions is at most $2\binom{6C}{2}+2$, since an arrangement of $m\geq 2$ great circles on a sphere induces at most $2(m2)+2$ 2(m2)+2 faces.

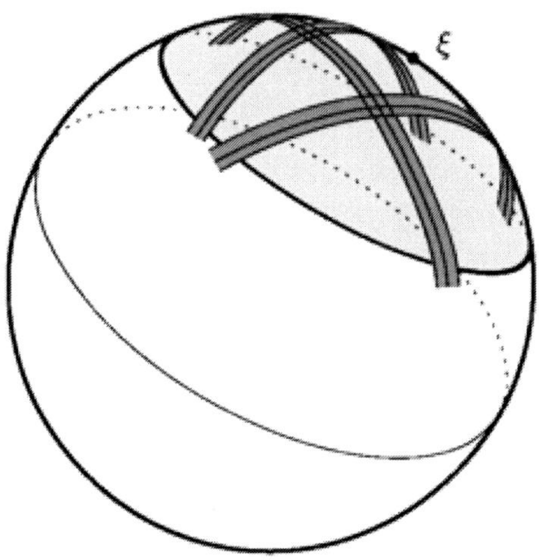

Figure. 2. The cap C_ξ (*lightly shaded*) being crossed by bands of half-width ε

Therefore the largest region has area at least

$$A := \frac{\text{area}(C_\xi) - 6C \cdot 4\pi \sin \varepsilon}{2\binom{6C}{2} + 2}.$$

A convex region of area $A<2\pi$ on the unit sphere always contains a spherical cap of radius at least $A/4$. Plugging in $\sin\varepsilon=148C$ sin⁡ε=148C, we conclude that there is a spherical cap of radius

$$\frac{1}{4} \cdot \frac{\text{area}(C_\xi) - 6C \cdot 4\pi \sin \varepsilon}{2\binom{6C}{2} + 2} = \frac{1}{4} \cdot \frac{\pi - 6C \cdot 4\pi \sin \varepsilon}{2\binom{6C}{2} + 2} = \frac{1}{4} \cdot \frac{\pi/2}{2\binom{6C}{2} + 2} = \frac{\pi}{16\binom{6C}{2} + 16} = \rho,$$

such that all directions in the cap satisfy conditions (P1) and (P2). By construction, Φ contains a direction in this cap. □

Now we are ready to prove Theorem 1 for the special case of constant-complexity polytopes.

Lemma 5

Let $0<\beta\leq 1/8$ and $C\geq 4$ be constants. Let P *be a set of n convex, β-fat polytopes in* $\mathbb{R}^3$, *each with at most C faces. Then*

$$\left\|\text{skel}\left(\bigcup \mathcal{P}\right)\right\| \leqslant K'_{\beta,C} \cdot \textstyle\sum_{P\in\mathcal{P}} \|\text{skel}(P)\| \cdot \alpha(n) \log^* n,$$

with $K'\ \beta,C$ Kβ,C′ *a constant depending only on β and C.*

Proof

Let Φ be the set of directions from Lemma 4. Put S:=skel(UP) S:=skel⁡(UP). Fix a direction $\phi\in\Phi$, and assume without loss of generality that ϕ is vertical. Let P ϕ ⊆P Pϕ⊆P be the set of all polytopes for which ϕ is β'-fattening, where β' is defined as in Lemma 4. Let $h(z)$ be the horizontal plane at height z. Let $c\ \phi$ (z) be the number of vertices of the two-dimensional union $h(z)$∩(UP ϕ) h(z)∩(UPϕ). Put $I_\phi := \int c_\phi(z)dz$ and $I := \sum_{\phi\in\Phi} I_\phi$. We show below that

1. (i) $\|S\| \leqslant 2I$, and
2. (ii) $I \leqslant K''_{\beta,C} \cdot \sum_{P\in\mathcal{P}}$ dia m$(P) \cdot \alpha(n) \log^* n$,

for an appropriate constant $K''\ \beta,C$ Kβ,C″. These will imply the lemma.

We begin with (i). Note that each edge e of S appears on $\mathrm{skel}(\bigcup \mathcal{P}_\phi)$ for some $\phi \in \Phi$ (possibly as part of a longer edge) where e makes an angle at most $\pi/3$ with ϕ—see properties (P1) and (P2). In particular, it contributes are least $\|e\|\cos(\pi/3)=\|e\|/2$ to I_ϕ and thus to I. Therefore, $\|S\|\leq 2I$, completing the proof of (i).

To argue (ii) we will show there is a constant $K'''_{\beta',C}$ such that

$$I_\phi = O(K'''_{\beta',C} \cdot \alpha(n) \log^* n) \cdot \sum_{P \in \mathcal{P}_\phi} \mathrm{diam}(P).$$

The desired statement then follows by summing over all $\phi \in \Phi$, and using $|\Phi| = O(C^4)$ and $\beta'=\Omega(\beta)$.

As we sweep the plane $h(z)$ over $\mathcal{P}_\phi$, it enters and leaves polytopes. We subdivide the range of values of z into maximal intervals $Z_1, Z_2, \ldots$ so that over each interval Z_j the plane $h(z)$ intersects the same subset $\mathcal{P}_j \subset \mathcal{P}_\phi$ of polytopes. We decompose I_ϕ into subintegrals I_j, each taken over one interval Z_j. For a fixed Z_j, the intersection of $h(z)$ with $\mathcal{P}_j$ yields a collection of β'-fat polygons with at most $C|\mathcal{P}_j|$ edges in total. Thus the union complexity is at most $f_{\beta'}(C|\mathcal{P}_j|)$ by Fact 3. In particular, $c_\phi(z)\leq f_{\beta'}(C|\mathcal{P}_j|)$ for these values of z and thus

$$\begin{aligned}
I_j &\leqslant f_{\beta'}(C|\mathcal{P}_j|) \cdot \|Z_j\| \\
&= \frac{f_{\beta'}(C|\mathcal{P}_j|)}{|\mathcal{P}_j|} \cdot |\mathcal{P}_j| \cdot \|Z_j\| \\
&\leqslant \frac{f_{\beta'}(Cn)}{n} \cdot |\mathcal{P}_j| \cdot \|Z_j\| \\
&= O(K'''_{\beta',C} \cdot \alpha(n) \log^* n) \cdot |\mathcal{P}_j| \cdot \|Z_j\|,
\end{aligned}$$

where $K'''_{\beta',C}$ is an appropriate constant. Hence,

$$I_\phi = \sum_j I_j = O(K'''_{\beta',C} \cdot \alpha(n) \log^* n) \cdot \sum_j |\mathcal{P}_j| \cdot \|Z_j\|.$$

Notice that each polytope $P \in \mathcal{P}_\phi$ contributes its z-span to the sum $\sum_j |\mathcal{P}_j|\cdot\|Z_j\|$. Hence, this sum is simply equal to the sum of the z-spans of all polytopes in $\mathcal{P}_\phi$. This sum is, in turn, at most $\sum_{P \in \mathcal{P}_\phi} \mathrm{diam}(P)$, as claimed.

Summing over all $\phi \in \Phi$, we obtain

$$\begin{aligned} I &= \sum_{\phi\in\Phi} I_\phi \\ &= \sum_{\phi\in\Phi} \left(O(K'''_{\beta',C} \cdot \alpha(n) \log^* n) \cdot \sum_{P\in\mathcal{P}_\phi} \mathrm{diam}(P) \right) \\ &= O(C^4 K'''_{\beta',C} \cdot \alpha(n) \log^* n \cdot \sum_{P\in\mathcal{P}} \mathrm{diam}(P)), \end{aligned}$$

which completes the proof of the lemma. □

The next lemma shows that the bound in Lemma 5 is close to the best possible in terms of its dependence on *n*.

Lemma 6

There exists a set Q *of n convex,* $\Theta(1)$*-fat polytopes in* $\mathbb{R}^3$*, each of constant complexity and unit diameter, with* $\|\mathrm{skel}(\cup Q)\| = \Omega(n\alpha(n))$

Proof

Let I I be a set of *n* segments in the plane whose lower envelope has complexity $\Theta(n\alpha(n))$ [14, 17]. We scale the plane vertically—that is, we "flatten" the construction—so that the envelope remains combinatorially the same in all directions within $\pi/4$ of vertical; see **Figure**. 3(i). Note that this also ensures that any segment makes an angle less than $\pi/4$ with the *x*-axis. By further scaling, we reduce the length of any segment in I to be smaller than some constant $0<\varepsilon<1-2\sqrt{}/2$. For every segment s∈I, we construct a quadrilateral K_s as the convex hull of *s* and a unit-length horizontal segment centered at a point lying half a unit above the midpoint of *s*—see **Figure**. 3(ii). The fact that *s* makes an angle of less than $\pi/4$ with the *x*-axis ensures that K_s is a quadrilateral. Clearly K_s is fat, and the condition that $\varepsilon<1-2/2$ implies that its diameter is determined by the unit-length horizontal segment. Observe that the two edges incident to *s* make an angle of less than $\pi/4$ with the vertical direction. Together with the property that the lower envelope of I I is combinatorially the same in all directions within $\pi/4$ from vertical, this implies that the lower envelope of I I appears on the union of the sets K_s. Now extruding each set K_s into a unit-height prism Q_s produces a set of *n* $\Theta(1)$-fat polytopes of constant complexity whose union contains $\Omega(n\alpha(n))$ unit-length edges, one for every vertex of the lower envelope of I. All polytopes have diameter exactly $2\sqrt{2}$; a further scaling by a factor 1/2 gives the desired result.

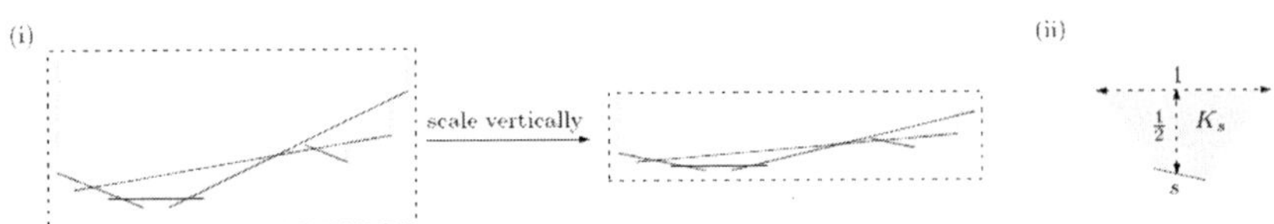

Figure. 3. (i) Flattening the construction. (ii) Making K_s out of *s*. (Not to scale.)

2.2 Non-constant-Complexity Polytopes

To extend the result of Lemma 5 to non-constant-complexity polytopes, we show that the boundary of a non-constant-complexity fat polytope can be covered using a linear number of constant-complexity fat polytopes. We start with a simple, well-known geometric fact.

Fact 7

For any compact convex set Γ in $\mathbb{R}^3$, there exists a simplex $\Delta(\Gamma) \subseteq \Gamma$ *with* $\mathrm{vol}(\Delta(\Gamma)) \geqslant \lambda \cdot \mathrm{vol}(\Gamma)$, *where* $\lambda = \frac{2}{3\sqrt{3}\pi}$.

Proof

Macbeath [9] has shown that the largest ratio between the volume of an object and the maximum volume of an inscribed simplex is achieved by a Euclidean ball. The result now follows from the fact that the ratio of the volume of a ball in $\mathbb{R}^3$ and its largest inscribed simplex (which is a regular tetrahedron) is 233 √

Lemma 8

Let P be a convex β-fat polytope with n_P vertices, and let f_{max} be the maximum number of vertices of any of its faces. Then there is a set T(*P*) T(P) *of constant-complexity convex polytopes such that*

1. *each polytope Q*∈T(*P*) Q∈T(P) *is Ω(β)-fat and contained in P,*

2. *the polytopes together cover the boundary of P,*

3. $\|\mathrm{skel}(P)\| \leqslant \sum_{Q\in T(P)} \|\mathrm{skel}(Q)\| = O(\|\mathrm{skel}(P)\| \cdot \log f_{\max})$, *and*

4. $|T(P)| = O(n_P)$.

Proof

First we triangulate each face *f* of *P*, as follows. Let *f* have *k* vertices, numbered in clockwise order as $v_0,\ldots,v_{k-1}$; the indices are treated modulo *k*. Now put in the diagonals $v_i v_{i+2}$ for $0 \le i \le k-2$ and *i* even (the total length of these diagonals is at most the perimeter of *f*) and recursively triangulate the part of *P* that still needs to be triangulated. The algorithm finishes after $\lceil \log_2 k \rceil$ rounds, and so the total edge length of all triangles in the triangulation is $O(\log k)=O(\log f_{\max})$ times the perimeter of f. The total number of triangles over all faces is obviously $O(n_P)$.

Let t be one such triangle. Put $d := \mathrm{diam}(t)$. Let σ be a cube centered at a point of t and having a point of t on its boundary, such that $t \subset \sigma$. The edge length of σ is at most $2d$. Put $P_t := P \cap \sigma$. Let B denote a largest inscribed ball of σ. Because P is β-fat, we have

$$\mathrm{vol}(P_t) \geqslant \mathrm{vol}(B \cap P_t) = \mathrm{vol}(B \cap P) \geqslant \beta \cdot \mathrm{vol}(B) = \beta \cdot \frac{\pi}{6} \cdot \mathrm{vol}(\sigma).$$

Define Q_t to be the convex hull of the union of t and $\Delta(P_t)$, with $\Delta(\cdot)$ as in Fact 7. The polytope Q_t is contained in P and has constant complexity, being the convex hull of at most seven points; also, t is contained in $\mathrm{bd}Q_t$. Moreover, $\|\mathrm{skel}(Q_t)\|=O(\text{perimeter of } t)$ ‖skel(Qt)‖=O(perimeter of t), since $d \leq \mathrm{diam}(Q_t) \leq 2d\sqrt{3}$ d≤diam(Qt)≤2d3. We claim that Q_t is $\Omega(\beta)$-fat. Indeed, we have

$$\mathrm{fatness}(Q_t) = \frac{\mathrm{vol}(Q_t)}{\frac{4}{3}\pi \cdot \mathrm{diam}^3(Q_t)} \geqslant \frac{\mathrm{vol}(\Delta(P_t))}{\frac{4}{3}\pi \cdot \mathrm{diam}^3(\sigma)} \geqslant \frac{\frac{2}{3\sqrt{3}\pi} \cdot \beta \cdot \frac{\pi}{6} \cdot \mathrm{vol}(\sigma)}{\frac{4}{3}\pi \cdot \mathrm{diam}^3(\sigma)} = \Omega(\beta).$$

.

Here the first equality follows from the fatness of a convex object o being determined by a ball whose radius is $\mathrm{diam}(o)$ [15, 16].

The lemma now follows by taking $\mathcal{T}(P) := \bigcup_t \{Q_t\}$, where the union is taken over all triangles t in the triangulations of the faces of P. □

We can now prove our main theorem.

Proof of Theorem 1

The lower bound follows from Lemma 6. To prove the upper bound, we first apply Lemma 8 to each of the polytopes $P \in \mathcal{P}$. Define $P \in \mathcal{P}$. Define $\mathcal{Q} := \bigcup_{P \in \mathcal{P}} \mathcal{T}(P)$. Then

- each element $Q \in \mathcal{Q}$ Q∈Q is a β'-fat polytope with at most C faces, for suitable constants $C \geq 4$ and $\beta' = \Omega(\beta)$;
- $|\mathcal{Q}| = O(\sum_{P \in \mathcal{P}} n_P) = O(n)$ |Q|=O(∑P∈PnP)=O(n); and
- $\bigcup \mathcal{Q} \subset \bigcup \mathcal{P}$ and $\mathrm{bd}(\bigcup \mathcal{P}) \subset \mathrm{bd}(\bigcup \mathcal{Q})$. which implies that $\|\mathrm{skel}(\bigcup \mathcal{P})\| \leqslant \|\mathrm{skel}(\bigcup \mathcal{Q})\|$.

Now we can apply Lemma 5 to the set $\mathcal{Q}$ Q to see that

$$\left\|\text{skel}\left(\bigcup \mathcal{P}\right)\right\|$$
$$\leq \left\|\text{skel}\left(\bigcup \mathcal{Q}\right)\right\|$$
$$\leq K'_{\beta',C} \cdot \sum_{Q\in\mathcal{Q}} \|\text{skel}(Q)\| \cdot \alpha(n)\log^* n \qquad \text{(by Lemma 5)}$$
$$\leq K'_{\beta',C} \cdot \sum_{P\in\mathcal{P}} \sum_{Q\in\mathcal{T}(P)} \|\text{skel}(Q)\| \cdot \alpha(n)\log^* n \qquad \text{(definition of } \mathcal{Q}\text{)}$$
$$\leq K'_{\beta',C} \cdot \sum_{P\in\mathcal{P}} O\left(\|\text{skel}(P)\| \cdot \log f_{\max}\right) \cdot \alpha(n)\log^* n \qquad \text{(by Lemma 8(iii))}$$

Since C is an absolute constant, we have

$$\|\text{skel}(\bigcup \mathcal{P})\| \leqslant K_\beta \cdot \sum_{P\in\mathcal{P}} \|\text{skel}(P)\| \cdot \log f_{\max} \cdot \alpha(n)\log^* n.$$
,

for some suitable K_β depending only on β. This finishes the proof of Theorem 1.

3 CONCLUSION

We have shown that the total length of the skeleton of the union of fat 3-dimensional polytopes is roughly linear in the sum of the skeleton lengths of the individual polytopes. More precisely, the skeleton length is $O(\alpha(n)\log^* n \cdot \log f_{\max})$ times the sum of the skeleton lengths of the individual polytopes, where $f_{\max}$ is the maximum complexity of any polytope facet. We also gave an example of n constant-complexity fat polytopes where the skeleton length is $O(\alpha(n))$ times the sum of the skeleton lengths of the individual polytopes. Hence, our bound is not far from the best possible. It would be interesting to know if the $\log f_{\max}$ factor in the upper bound of Theorem 1 is necessary.

Another question is what happens when the objects are curved. As an example, consider the skeleton of a set $B=\{b_1,\ldots,b_n\}$ of balls in $\mathbb{R}^3$. Since the cross section of a ball with a plane is a disk, and the union complexity of m disks is $O(m)$, our technique can be used to show that $\|\text{skel}(\bigcup \mathcal{B})\| = O(\sum_i \text{diam}(b_i))$. (Indeed, if we sweep three times in mutually orthogonal directions, then for any point q on a skeleton edge there will be a sweep such that the direction of the edge at q makes a sufficiently large angle with the sweep plane.) It would be interesting to generalize this result to other classes of fat curved objects.

ACKNOWLEDGEMENTS

The authors wish to thank Erwin Lutwak and Alfredo Hubard for useful pointers and anonymous referees for a number of helpful suggestions.

Work by B.A. has been partially supported by a grant from the U.S.–Israel Binational Science Foundation, by NSA MSP Grant H98230-06-1-0016, and by NSF Grant CCF-08-30691.

This research was supported by the Netherlands' Organisation for Scientific Research (NWO) under project no. 639.023.301.

REFERENCES

1. Agarwal, P.K., Pach, J., Sharir, M.: State of the union (of geometric objects): A review. In: Goodman, J., Pach, J., Pollack, R. (eds.) Surveys on Discrete and Computational Geometry: Twenty Years Later, pp. 9–48. Am. Math. Soc., Providence (2008)
2. Aronov, B., de Berg, M., Ezra, E., Sharir, M.: Improved bound for the union of locally fat objects in the plane. Manuscript
3. Boissonnat, J.-D., Sharir, M., Tagansky, B., Yvinec, M.: Voronoi diagrams in higher dimensions under certain polyhedral distance functions. Discrete Comput. Geom. **19**, 485–519 (1998)
4. 4.Bonnesen, T., Fenchel, W.: Theorie der konvexen Körper. Springer, Berlin (1974)
5. de Berg, M.: Improved bounds on the union complexity of fat objects. Discrete Comput. Geom. **40**, 127–140 (2008)
6. de Berg, M., Gray, C.: Vertical ray shooting and computing depth orders for fat objects. SIAM J. Comput. **38**, 257–275 (2008)
7. .Ezra, E., Sharir, M.: Almost tight bound for the union of fat tetrahedra in three dimensions. J. ACM **57**(1), 2 (2009)
8. Ezra, E., Aronov, B., Sharir, M.: Improved bound for the union of fat triangles. In: Proc. Annual ACM-SIAM Symp. Discr. Algorithms, pp. 1778–1785 (2011)
9. Macbeath, A.M.: An extremal property of the hypersphere. Proc. Camb. Philos. Soc. **47**, 245–247 (1951)
10. 10Matoušek, J., Pach, J., Sharir, M., Sifrony, S., Welzl, E.: Fat triangles determine linearly many holes. SIAM J. Comput. **23**, 154–169 (1994)
11. Pach, J., Tardos, G.: On the boundary complexity of the union of fat triangles. SIAM J. Comput. **31**, 1745–1760 (2002)
12. Pach, J., Safruti, I., Sharir, M.: The union of congruent cubes in three dimensions. Discrete Comput. Geom. **30**, 133–160 (2003)

13. Pettie, S.: On the structure and composition of forbidden sequences, with geometric applications. In: Proc. 27th Annu. Symp. Comput. Geometry, pp. 370–379 (2011)

14. Sharir, M., Agarwal, P.K.: Davenport–Schinzel Sequences and Their Geometric Applications. Cambridge University Press, New York (1996)

15. van der Stappen, A.F.: Motion planning amidst fat obstacles. Ph.D. thesis, Utrecht University, Utrecht, the Netherlands (1994)

16. van der Stappen, A.F., Halperin, D., Overmars, M.H.: The complexity of the free space for a robot moving amidst fat obstacles. Comput. Geom., Theory Appl. **3**, 353–373 (1993)

17. Wiernik, A., Sharir, M.: Planar realization of nonlinear Davenport–Schinzel sequences by segments. Discrete Comput. Geom. **3**, 15–47 (1988)

CHAPTER 9

Sphere anDot Product Representations of Graphs

Ross J. Kang, Durham University, Tobias Müller

Durham University

Centrum Wiskunde & Informatica

ABSTRACT

A graph G is a k-sphere graph if there are k-dimensional real vectors $v_1,\ldots,v_n$ such that $ij \in E(G)$ if and only if the distance between v_i and v_j is at most 1. A graph G is a k-dot product graph if there are k-dimensional real vectors $v_1,\ldots,v_n$ such that $ij \in E(G)$ if and only if the dot product of v_i and v_j is at least 1.

By relating these two geometric graph constructions to oriented k-hyperplane arrangements, we prove that the problems of deciding, given a graph G, whether G is a k-sphere or a k-dot product graph are NP-hard for all $k>1$. In the former case, this proves a conjecture of Breu and Kirkpatrick (Comput. Geom. 9:3–24, 1998). In the latter, this answers a question of Fiduccia et al. (Discrete Math. 181:113–138, 1998).

Furthermore, motivated by the question of whether these two recognition problems are in NP, as well as by the implicit graph conjecture, we demonstrate that, for all $k>1$, there exist k-sphere graphs and k-dot product graphs such that each representation in k-dimensional real vectors needs at least an exponential number of bits to be stored in the memory of a computer. On the other hand, we show that exponentially many bits are always enough. This resolves a question of Spinrad (Efficient Graph Representations, 2003).

Keywords: Geometric representations of graphs Sphericity Dot product dimension

1 INTRODUCTION AND STATEMENT OF RESULTS

A graph G is a k-sphere graph if there are vectors $v_1,\ldots,v_n \in \mathbb{R}^k$ such that $\|v_i - v_j\| \leq 1$ if and only if $ij \in E(G)$. Alternatively, a k-sphere graph can be seen as the intersection graph of equal radius spheres in k-dimensional space; that is, we can represent each vertex i by a sphere $S_i \subseteq \mathbb{R}^k$ of radius 1 in such a way that $ij \in E(G)$ if and only if $S_i \cap S_j \neq \emptyset$. The *sphericity* sph($G$) of a graph G is the least k such that G is a k-sphere graph. (Every graph has finite sphericity [10].) Sphericity was first introduced by Havel [7] in his Ph.D. thesis to study the geometry of molecules. Graphs of sphericity one are also called *unit interval graphs*, and graphs of sphericity at most two are called *unit disk graphs*. Unit disk graphs have received considerable attention over the last two decades, in part because of their relevance to applications in wireless networks.

A *k-dot product graph* is a graph G that can be represented by vectors $v_1,\ldots,v_n \in \mathbb{R}^k$ such that $ij \in E(G)$ if and only if $v_i^T v_j \geq 1$ The *dot product dimension* dpd(G) of a graph G is the least k such that G is a k-dot product graph. (Every graph has finite dot product dimension [6].) Dot product representations of graphs were first introduced by Reiterman et al. [16–18] and (independently, it seems) by Fiduccia et al. [6]. Here, it should be mentioned that Reiterman et al. used an arbitrary threshold $t \in \mathbb{R}$ in their definition (i.e. $ij \in E(G)$ if and only if $v_i^T v_j \geq t$ viTvj≥t). It is however straightforward to adapt our proofs to this definition and we fix $t=1$ for convenience. Graphs of dot product one are precisely the class of graphs that have at most two nontrivial connected components each of which is a threshold graph, cf. [6]. The class of threshold graphs has been studied quite extensively, cf. [11].

In this paper, we will consider two related types of questions. First, we study the algorithmic decision problems of recognizing k-sphere graphs and k-dot product graphs. Second, we analyse the number of bits needed to store the coordinates of the vectors $v_1,\ldots,v_n \in \mathbb{R}^k$ in a k-sphere or a k-dot product representation.

By k-SPHERICITY, we mean the problem of deciding, given a graph G as input, whether G has sphericity at most k. Put in different words, k-SPHERICITY is the recognition problem for k-sphere graphs. It is known that the $k=1$ version of this problem can be solved in linear time [9]. On the other hand, Breu and Kirkpatrick [4] proved that 2-SPHERICITY is NP-hard. Breu and Kirkpatrick used a rather complicated reduction from SATISFIABILITY, which crucially relies on a special embedding of instances of SATISFIABILITY on a 2-dimensional grid. They briefly sketched how a similar embedding on a 3-dimensional grid can also be obtained, showing NP-hardness of 3-sphericity, but were unable to extend this further and they conjectured that k-SPHERICITY is NP-hard for all remaining k. We prove their conjecture, by providing a simpler NP-hardness proof for 2-SPHERICITY that also extends to all higher dimensions.

Theorem 1

k-SPHERICITY is NP-hard for all $k>1$.

The proof of Theorem 1 builds upon techniques that appeared earlier in [13].

The decision problem k-DOT PRODUCT asks whether a given input graph has dot product dimension at most k. For $k=1$ it follows from a result of Fiduccia et al. (Theorem 20 in [6]) and a result of Chvátal and Hammer that this is solvable in linear time. Fiduccia et al. [6] asked whether this recognition problem is NP-hard for other values of k, and we affirm this.

Theorem 2

k-DOT PRODUCT is NP-hard for all $k>1$.

Rather than a reduction from (some variant of) SATISFIABILITY, the proofs of Theorems 1 and 2 follow a reduction from the so-called simple stretchability problem, to be defined in the next section.

Having determined that the decision problems of recognizing k-sphere graphs and k-dot product graphs are NP-hard, one might wonder whether these decision problems are members of the decision class NP. For this, we need a "polynomial certificate". That is, for each graph that is a k-sphere (resp. k-dot product) graph, there is a proof of this fact that can be checked by an algorithm in polynomial time. An obvious candidate for such a certificate is the set of vectors $v_1,\ldots,v_n$ of a k-sphere (resp. k-dot product) representation. For this to be a good certificate, we would however need to guarantee that $v_1,\ldots,v_n$ can be stored using polynomially many bits.

Spinrad [22] also asked how many bits are needed to store the coordinates of a k-dot product graph, motivated by the so-called *implicit graph conjecture* of Kannan et al. [8]. This conjecture asserts that, for every hereditary graph class C (i.e. C C is closed under taking induced subgraphs) such that the number of graphs on n vertices in C is $2^{O(n\ln n)}$, there exists a "labelling scheme" that encodes graphs in the class by assigning bit strings of length $O(\ln n)$ to the vertices in such a way that the adjacency of two vertices u,v can be tested by examining only the labels of u and v. It is easily shown that the number of k-sphere graphs (resp. k-dot product graphs) on n vertices is $2^{O(n\ln n)}$ by Warren's Theorem, cf. [22]. A natural candidate for the labelling scheme would be to use the vectors $v_1,\ldots,v_n$ of some k-sphere (resp. k-dot product) representation as the labels. For this to work, we would however need to guarantee that each of $v_1,\ldots,v_n$ can be stored using only $O(\ln n)$ bits.

The bit size of a natural number $n\in\mathbb{N}$ is the number of symbols needed to write it in binary notation. It satisfies

$$\text{size}(n) := \lceil \log_2(n+1) \rceil. \tag{1}$$

The bit size of an integer $z\in\mathbb{Z}$ is $\text{size}(z)=1+\text{size}(|z|)$, as we need an extra bit for the sign. In this paper, we shall use the convention that a rational number is stored as a pair of integers (the denominator and numerator) that are relatively prime and these integers are stored in the binary number format. Following Basu et al. [2], we define the size of a rational number $q=a/b\in\mathbb{Q}$ with $a,b\in\mathbb{Z}$ relatively prime as

$$\operatorname{size}(q) := \operatorname{size}(a) + \operatorname{size}(b). \tag{2}$$

For a vector $v \in \mathbb{Q}^k$, we set $\operatorname{size}(v) := \sum_{i=1}^{k} \operatorname{size}(v_i)$.

We should remark that several other notions of the size of a rational number exist in the literature.

We are interested in the following quantities for G a k-sphere graph (resp. a k-dot product graph):

$$f_k(G) := \min_{\substack{v: V(G) \to \mathbb{Q}^k, \\ v \text{ is a } k\text{-sphere} \\ \text{representation of } G}} \sum_{t \in V(G)} \operatorname{size}(v(t)),$$

$$g_k(G) := \min_{\substack{v: V(G) \to \mathbb{Q}^k, \\ v \text{ is a } k\text{-dot product} \\ \text{representation of } G}} \sum_{t \in V(G)} \operatorname{size}(v(t)).$$

It may not be immediately clear to the reader that $f_k(G), g_k(G)$ are well-defined, since this requires that every k-sphere or k-dot product graph has a representation with all coordinates of all vectors rational. That such a rational representation always exists follows from (the proofs of) the upper bounds in Theorems 3 and 4 below. Let us set

$$f_k(n) := \max_{\substack{|V(G)|=n, \\ G \text{ is a } k\text{-sphere graph}}} f_k(G),$$

and

$$g_k(n) := \max_{\substack{|V(G)|=n, \\ G \text{ is a } k\text{-dot product graph}}} g_k(G).$$

We will show that there are k-sphere graphs that require exponentially many bits, but exponentially many bits are always enough.

Theorem 3

For every $k>1$, *there exist* $c_1, c_2 > 0$ *such that*

$$2^{c_1 n} \le f_k(n) \le 2^{c_2 n}.$$

Here is should be mentioned that the special case when k=2 of Theorem 3 was previously proved by McDiarmid and the second author [12], and a slightly weaker lower bound for the case k=2 of $f2\,(n) \geq 2\;\Omega(n\;\sqrt{\;})$ f2(n)≥2Ω(n) was previously proved in [13], the conference version of [12].

We also show that the analogous result holds for k-dot product graphs.

Theorem 4

For every k>1, there exist c_1, c_2>0 such that

$$2^{c_1 n} \leq g_k(n) \leq 2^{c_2 n}.$$

Theorems 3 and 4 show that the vectors of a k-sphere representation or of a k-dot product representation cannot be used as a labelling scheme to prove a special case of the implicit graph conjecture. They also cannot serve as certificates to prove membership in NP of the recognition problems.

2 Preliminaries

A *line arrangement* is a system L:=(ℓ 1 ,...,ℓ n) L:=(ℓ1,...,ℓn) of lines in the plane. If every two lines intersect and no point lies on more than two lines, then we call L L a *simple line arrangement* (or, also, a *uniform line arrangement*). See **Figure**. 1 for a depiction of a non-simple and a simple line arrangement.

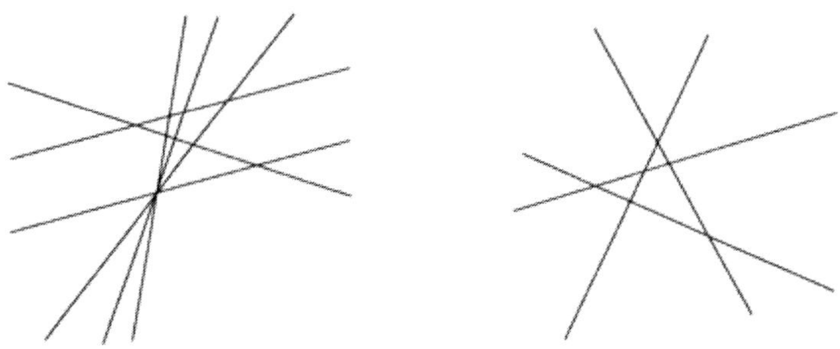

Figure. 1. A non-simple and a simple line arrangement

A line ℓ divides $\mathbb{R}^2\backslash\ell$ into two pieces. In an *orientation* of ℓ, we distinguish between these two pieces by (arbitrarily) calling one of them, ℓ^-, the "negative side" and the other, ℓ^+, the "positive side". An *oriented line arrangement* is a line arrangement for which all the lines have been given an orientation.

For convenience, from now on we shall only work with oriented line arrangements. The *sign vector* of a point $p\in\mathbb{R}^2$ with respect to an oriented line arrangement L=(ℓ 1 ,...,ℓ n) L=(ℓ1,...,ℓn) is a vector $\sigma(p)\in\{-,0,+\}^n$ defined as follows:

$$(\sigma(p))_i = \begin{cases} - & \text{if } p \in \ell_i^-, \\ 0 & \text{if } p \in \ell_i, \\ + & \text{if } p \in \ell_i^+. \end{cases}$$

The *combinatorial description* D(L) D(L) of L L is the set of all sign vectors

$$\mathcal{D}(\mathcal{L}) := \{\sigma(p) : p \in \mathbb{R}^2\}.$$

The set D(L) D(L) is almost the same thing as the set of covectors of an oriented matroid and in fact it determines the oriented matroid associated with L L. (Consult the monograph on oriented matroids by Björner et al. [3] for more details.) It should be mentioned that various alternative combinatorial descriptions of a line arrangement are available, such as local sequences, allowable sequences and wiring diagrams (cf. [5]).

Each connected component of $\mathbb{R}^2 \setminus (\ell_1 \cup \cdots \cup \ell_n)$ is called a *cell* or a *region*. All points in the same cell have the same sign vector, which does not have 0 as a coordinate. A sign vector with two or more zeroes corresponds to an intersection point between two or more lines, and a sign vector with exactly one zero corresponds to a line segment.

If $\mathcal{D}(\mathcal{L}) = \mathcal{D}(\mathcal{L}')$, then we say that L L and L ′ L′ are *isomorphic*. We will use a number of relatively straightforward observations about (oriented) line arrangements in our proofs below. For completeness, we include their proofs in the Appendix.

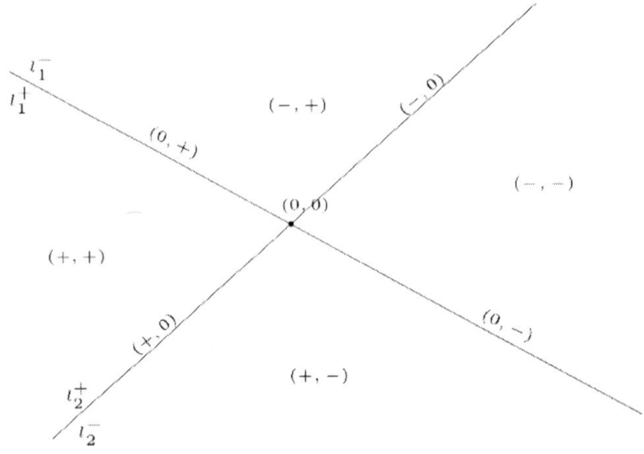

Lemma 5

Let L *be a line arrangement. Then there exists* $\sigma \in \mathcal{D}(\mathcal{L}) \cap \{-,+\}^n$ *such that* $-\sigma \in \mathcal{D}(\mathcal{L}) \cap \{-,+\}^n$.

Lemma 6

A line arrangement on n lines has at most $1+\binom{n+1}{2}$ *cells, and it is simple if and only if it has exactly* $1+\binom{n+1}{2}$ *cells.*

Lemma 7

Let L L *be a simple oriented line arrangement, and let* L ' *be an oriented line arrangement with the same number of lines. Then* L *and* L ' *are isomorphic if and only if* $\mathcal{D}(\mathcal{L}) \cap \{-,+\}^n = \mathcal{D}(\mathcal{L}') \cap \{-,+\}^n$.

Similarly to a line arrangement in the plane, a *k*-hyperplane arrangement H=($h_1,\ldots,h_n$) is a system of (affine) hyperplanes in $\mathbb{R}^k$. We define a cell, an orientation, a sign vector and the combinatorial description in precisely the same way as for lines.

Recall that an *affine transformation* $T:\mathbb{R}^k \rightarrow \mathbb{R}^k$ is a linear transformation followed by a translation. Put differently, it is a map of the form $x \mapsto Ax+b$, where A is a nonsingular $k \times k$ matrix, and $b \in \mathbb{R}^k$ is a (constant) vector. We will need the following observation in the sequel. The straightforward proof can again be found in the Appendix.

Lemma 8

Let H H *be an oriented k-hyperplane arrangement, and let* $T:\mathbb{R}^k \rightarrow \mathbb{R}^k$ *be an affine transformation. Define another oriented k-hyperplane arrangement by setting*

$$h_i' := T[h_i], \qquad (h_i')^- := T[h_i^-], \qquad (h_i')^+ := T[h_i^+].$$

Then H *and* H ' *are isomorphic.*

STRETCHABILITY is the problem of deciding, given a set $S \subseteq \{-,0,+\}^n$, whether S is the combinatorial description of some oriented line arrangement of n lines. We can also define an orientation and a combinatorial description for an arrangement C=($c_1,\ldots,c_n$) of continuous curves (that satisfy some regularity conditions) in the same way. The name "stretchability" now comes from the idea that we may ask, for an oriented system of curves in the plane, whether there is a line arrangement with the same combinatorial description. In this case, we may imagine the curves "stretching" into lines.

Similarly, SIMPLE STRETCHABILITY is the problem of deciding, given a set $S \subseteq \{-,0,+\}^n$, whether S is the combinatorial description of some oriented simple arrangement of n lines.

Theorem 9

(Shor [21])

SIMPLE STRETCHABILITY is NP-hard.

This theorem is also a straightforward corollary of a deep topological result by Mnëv [14]. Shor's proof is more direct. It reduces a SAT-variant to stretchability using Pappus' anDesargues' theorems, cf. Chap. 8 of [3].

3 THE PROOFS

For the NP-hardness proofs, it will be useful to define an intermediate decision problem. For $k>1$ and $S \subseteq \{-,+\}^n$ with $(-,\ldots,-),(+,\ldots,+) \in S$, we say that S is *k-realizable* if there exists an oriented k-hyperplane arrangement H H with $S \subseteq D(H)$. The condition that $(-,\ldots,-),(+,\ldots,+) \in S$ is useful for technical reasons later on in the proofs of Theorems 1 and 2. For $k>1$, we let k-REALIZABILITY denote the algorithmic problem of deciding, given a set $S \subseteq \{-,+\}^n$ as input, whether S is k-realizable.

Theorem 10

k-REALIZABILITY is NP-hard for all $k>1$.

The proof of Theorem 10 relies on the following observation:

Lemma 11

Let $S \subseteq \{-,+\}^n$ with $(-,\ldots,-),(+,\ldots,+) \in S$ and $k>1$ be arbitrary and set $S' := S \times \{-,+\}$. Then S is k-realizable if and only if S' is $(k+1)$-realizable.

Proof

Let k, S, S' be as in the statement of the lemma. Let us first suppose that there exists an oriented hyperplane arrangement H=(h 1 ,…,h n) H=(h1,…,hn) in $\mathbb{R}^k$ such that $S \subseteq D(H)$ S⊆D(H). Define a new hyperplane arrangement H ′ H′ by

$$h'_i = h_i \times \mathbb{R}, \qquad (h'_i)^- = h_i^- \times \mathbb{R}, \qquad (h'_i)^+ = h_i^+ \times \mathbb{R},$$

for $i=1,\ldots,n$, and

$$h'_{n+1} = \mathbb{R}^k \times \{0\}, \qquad (h'_{n+1})^- = \mathbb{R}^k \times \{x : x < 0\},$$
$$(h'_{n+1})^+ = \mathbb{R}^k \times \{x : x > 0\}. \tag{3}$$

For each $s \in S$, there exists $p \in \mathbb{R}^k$ such that $\sigma_{\mathrm{H}}(p) = s$. If we set $p_1 = (p,-1)$ then clearly $\sigma_{\mathrm{H}'}(p_1) = (s,-)$ σH′(p1)=(s,−) and $\sigma_{\mathrm{H}'}(p_2) = (s,+)$. Thus, $S' \subseteq \mathrm{D}(\mathrm{H}')$, as required.

Now, suppose that $S' \subseteq \mathrm{D}(\mathrm{H}')$ for some $(k+1)$-hyperplane arrangement H' H′. By Lemma 8, we can also assume without loss of generality that (3) holds. Let us define a k-hyperplane arrangement $\mathrm{H} = (h_1, \ldots, h_n)$ H=(h1,…,hn) by identifying $\mathbb{R}^k$ with $\mathbb{R}^k \times \{0\}$ in the obvious way and setting

$$h_i := h'_i \cap \mathbb{R}^k, \qquad h_i^- := (h'_i)^- \cap \mathbb{R}^k, \qquad h_i^+ := (h'_i)^+ \cap \mathbb{R}^k. \tag{4}$$

It is not immediately clear that each h_i is a k-hyperplane, since we might have $h_i = \emptyset$, in case h'_i hi′ and h'_{n+1} hn+1′ happen to be parallel, or $h_i = \mathbb{R}^k$, in case h'_i hi′ and h'_{n+1} hn+1′ coincide. However, if none of these two cases holds, then $\dim_{\mathbb{R}^{k+1}} (h'_i \cap h'_{n+1}) = k - 1$ so that h_i is a well-defined hyperplane in $\mathbb{R}^k$. We shall see soon that this is in fact the case for all $i = 1, \ldots, n$.

Let $s \in S$ be arbitrary, and let us write $s_1 = (s,-), s_2 = (s,+)$. Pick $p_1, p_2 \in \mathbb{R}^{k+1}$ such that $\sigma_{\mathrm{H}'}(p_i) = s_i$. Then $p_1 \in (h'_{n+1})^-$ and $p_2 \in (h'_{n+1})^+$, so that the segment $[p_1, p_2]$ must intersect $h'_{n+1} = \mathrm{R}^k \times \{0\}$ at some point $p' = (p,0)$. It is clear that $\sigma_{\mathrm{H}'}(p') = (s,0)$ and hence $\sigma_{\mathrm{H}}(p) = s$.

It remains for us to show that the h_i are well-defined hyperplanes in $\mathbb{R}^k$. To this end, recall that $(-,\ldots,-),(+,\ldots,+) \in S$. Let $p,q \in \mathbb{R}^k$ be points with $\sigma_{\mathrm{H}}(p) = (-,\ldots,-)$ σH(p)=(−,…,−), and let ℓ be the line between p,q. Then $\ell \subseteq \mathrm{R}^k = h'_{n+1}$ ℓ⊆Rk=hn+1′ and ℓ hits all of h_i^- hi−, h_i, h_i^+ hi+ for each $i = 1, \ldots, n$. Hence, we cannot have $h_i = \emptyset$ or $h_i = \mathbb{R}^k$ so that h_i is a well-defined k-hyperplane.

Proof of Theorem 10

We shall start by considering the case when $k=2$. For $k=2$, we shall use a reduction from SIMPLE STRETCHABILITY. Let $D \subseteq \{-,0,+\}^n$ be some set of sign vectors. We need to construct, in polynomial time, a set $S \subseteq \{-,+\}^n$ such that S is 2-realizable if and only if the answer to SIMPLE STRETCHABILITY with input D is "yes".

We can assume that $|D \cap \{-,+\}^n| = 1 + \binom{n+1}{2}$. This is because, if $|D \cap \{-,+\}^n| \neq 1 + \binom{n+1}{2}$, then the answer to the SIMPLE STRETCHABILITY-instance is "no" by Lemma 6. We can certainly determine if $|D \cap \{-,+\}^n| \neq 1 + \binom{n+1}{2}$ in polynomial time, and if so we can simply output a set $S \subseteq \{-,+\}^n$ which we know is not 2-realizable. Similarly, we can assume that there is some $s \in D \cap \{-,+\}^n$ such that $-s \in D$. (We can check for the existence of such a pair in polynomial time, and non-existence of such a pair would imply that the answer to the SIMPLE STRETCHABILITY-instance is "no" by Lemma 5.)

Now, let us pick such a pair $s,-s \in D \cap \{-,+\}^n$ and let $f:\{-,0,+\}^n \to \{-,0,+\}^n$ denote the map that changes + to − and changes − to + in all coordinates i where $s_i = +$ and leaves all other coordinates unchanged. Let us set $D' := \{f(t) : t \in D\}$. Note that $f(s)=(-,\ldots,-)$ and $f(-s)=(+,\ldots,+)$. If $D=\mathrm{D}(\mathrm{L})$ for some line arrangement, then D' corresponds to a "reorientation" (i.e. switching the roles of h^-_i hi− and h^+_i hi+ for some indices i). Thus, $D=\mathrm{D}(\mathrm{L})$ for some simple oriented line arrangement L L if and only if $D'=\mathrm{D}(\mathrm{L}')$ for some simple oriented line arrangement L'. (D' is constructed from D to ensure the technical promise that $(-,\ldots,-),(+,\ldots,+) \in S$ given earlier.)

We now set $S = D' \cap \{-,+\}^n$. We claim that S is 2-realizable if and only if $D=\mathrm{D}(\mathrm{L})$ for some simple oriented line arrangement L L.

To see this, first note that if $D=\mathrm{D}(\mathrm{L})$ for some simple L L, then S is 2-realizable by construction. Let us thus assume that S is 2-realizable. By Lemma 6, there is a *simple* L L such that $S \subseteq \mathrm{D}(\mathrm{L})$. Lemma 7 now implies that $\mathrm{D}(\mathrm{L})=D'$. But then we also have that D is the combinatorial description of some simple oriented line arrangement. Thus, S is 2-realizable if and only if D is the combinatorial description of a simple oriented line arrangement, as claimed. Since all the steps above to produce S from D can clearly be implemented in polynomial time, this proves the theorem for $k=2$.

Let us now consider the case $k>2$, and suppose that we have already shown that ($k-1$)-REALIZABILITY is NP-hard. Since it is clear that we can construct $S' = S \times \{-,+\}$ in polynomial time from S, it is a direct consequence of Lemma 11 that k-REALIZABILITY is also NP-hard. □

To prove Theorems 1 and 2, we will also need the following technical lemma.

Lemma 12

Let $\varepsilon>0$, $K>0$ *be arbitrary, and let* H H *be an oriented hyperplane arrangement with* $(-,\ldots,-),(+,\ldots,+) \in \mathrm{D}(\mathrm{H})$ (−,…,−),(+,…,+)∈D(H). *Then there exists* H' H′ *with the following properties.*

(i) (i)$\mathrm{D}(\mathrm{H}')=\mathrm{D}(\mathrm{H})$ D(H′)=D(H).

(ii) (ii)*The origin* 0 *belongs to the* $(-,\ldots,-)$*-cell of* H' H′.

(iii) (iii)*For each oriented hyperplane* h'_i hi′ *of* H' H′, *we can write*

(iv) $$h'_i = \{z : v_i^T z = c_i\}, \qquad h_i'^+ = \{z : v_i^T z < c_i\}, \qquad h_i'^+ = \{z : v_i^T z > c_i\},$$

(v) *with* $\|v_i - e_1\| < \varepsilon$ (*where* $e_1=(1,0,\ldots,0)$).

(vi) (iv)*Every cell of* H ′ H′ *intersects the ball* $B((K,0,\ldots,0)^T,\varepsilon)$.

Proof

The proof is by means of repeated applications of Lemma 8. By using a suitable scaling $z\mapsto\lambda z$ if needed, we can ensure that every cell of H H intersects the ball $B(0,\varepsilon/2)$ of radius $\varepsilon/2$ around the origin. In particular, there are points $p,q\in B(0,\varepsilon/2)$ such that $\sigma(q)=(-,\ldots,-)$, $\sigma(p)=(+,\ldots,+)$. Let ℓ denote the line through p and q. Notice that all the hyperplanes $h_1,\ldots,h_n$ must intersect ℓ in the open segment (p,q).

Let r be the connected component of $\ell\setminus(p,q)$ that contains q. Thus, r is a "ray" emanating from q. All points of r must have sign vector $(-,\ldots,-)$ (since none of the hyperplanes $h_1,\ldots,h_n$ hits r). By applying a suitable translation followed by a rotation, we can map q to the origin 0 and r to the nonnegative e_1-axis $\{(x,0,\ldots,0):x\leq 0\}$. Notice that now, since we have translated over a distance of $\|q\|<\varepsilon/2$, every cell intersects $B(0,\varepsilon)$. Also, notice that the positive e_1-axis intersects each of the hyperplanes $h_1,\ldots,h_n$.

Next, we will rescale only the e_1-axis to make all hyperplanes have normals that are close to the e_1-vector. To see that this can be done, let $v_1,\ldots,v_n$ be vectors and $c_1,\ldots,c_n$ be constants such that

$$h_i=\{z:v_i^Tz=c_i\},\qquad h_i^-=\{z:v_i^Tz<c_i\},\qquad h_i^+=\{z:v_i^Tz>c_i\}.$$

Note that, since $0\in h_i^-$ 0_∈hi−, we have $c_i>0$ for all i. Similarly, since $(x,0,\ldots,0)\in h_i^-$ (x,0,…,0)∈hi− for all $x\leq 0$, we must then also have $(v_i)_1\geq 0$. Also, note that, since e_1 is not parallel to h_i (recall that h_i intersects the positive e_1-axis), we must have $(v_i)_1\neq 0$ for all i. Now, let T denote the linear transformation $T:(z_1,\ldots,z_n)\mapsto(\lambda z_1,z_2,\ldots,z_n)$, for some $\lambda>0$. Then we obtain

$$\begin{aligned} h_i' &= \{T(z):v_i^Tz=c_i\} \\ &= \{z:\lambda^{-1}(v_i)_1z_1+(v_i)_2z_2+\cdots+(v_i)_nz_n=c_i\}. \end{aligned}$$

Let us write

$$w_i:=\left(\lambda^{-1}(v_i)_1,(v_i)_2,\ldots,(v_i)_k\right)^T,\qquad u_i:=w_i/\|w_i\|,\qquad d_i:=c_i/\|w_i\|.$$

Clearly, $h'_i=\{z:u_i^Tz=d_i\}$ hi′={z:uiTz=di}. Now, note that $u_i\to e_1$ as $\lambda\searrow 0$. Thus, if we take λ sufficiently small, then we obtain $\|u_i-e_1\|<\varepsilon$ for all i. Notice that, since we have merely "scaleDown" the e_1-coordinate of all points in $\mathbb{R}^k$, it still holds that every cell intersects $B(0,\varepsilon)$ and the nonpositive e_1-axis belongs to the all-minus cell.

Finally, if we apply the translation $z \mapsto z+Ke_1$, then it is clear that the normals of the hyperplanes do not change and the ball $B(0,\varepsilon)$ is mapped into $B((K,0,\ldots,0)^T,\varepsilon)$. Hence, every cell now hits $B((K,0,\ldots,0)^T,\varepsilon)$. Moreover, the origin still lies in the cell corresponding to $(-,\ldots,-)$, since the nonpositive e_1-axis was mapped onto $\{(x,0,\ldots,0):x\leq K\}$.

Proof of Theorem 1

We will use a reduction from k-REALIZABILITY. Thus, given a set $S\subseteq\{-,+\}^n$ with $(-,\ldots,-),(+,\ldots,+)\in S$, we will construct (in polynomial time) a graph G such that G is a k-sphere graph if and only if $S\subseteq D(H)$ S⊆D(H) for some oriented k-hyperplane arrangement H H.

Given $S\subseteq\{-,+\}^n$, let $G=G(S)$ be a graph on $2n+|S|$ vertices which is defined as follows. Let $V(G)=A\cup B\cup C$, where $A=\{a_1,\ldots,a_n\}$, $B=\{b_1,\ldots,b_n\}$ and $C=\{c_\sigma:\sigma\in S\}$. The edge set of G is as follows:

- $a_i a_j, b_i b_j \in E(G)$ for all $1\leq i<j\leq n$;
- $a_i b_j \notin E(G)$ for all $1\leq i,j\leq n$;
- $c_\sigma c_\tau \in E(G)$ for all $\sigma\neq\tau\in S$;
- $a_i c_\sigma \in E(G)$ if and only if $\sigma_i=-$;
- $b_i c_\sigma \in E(G)$ if and only if $\sigma_i=+$.

It should be clear that $G(S)$ can be constructed in polynomial time. It remains to show that G is a k-sphere graph if and only if there exists a k-hyperplane arrangement H H such that $S\subseteq D(H)$ S⊆D(H).

First, suppose that G can be realized as a k-sphere graph. That is, there is a function $v:V(G)\to\mathbb{R}^k$ such that $\|v(s)-v(t)\|\leq 1$ if and only if $st\in E(G)$. Let us define a hyperplane arrangement H H as follows:

$$\begin{aligned} h_i &:= \left\{z : \left\|z - v(a_i)\right\| = \left\|z - v(b_i)\right\|\right\}, \\ h_i^- &:= \left\{z : \left\|z - v(a_i)\right\| < \left\|z - v(b_i)\right\|\right\}, \\ h_i^+ &:= \left\{z : \left\|z - v(a_i)\right\| > \left\|z - v(b_i)\right\|\right\}. \end{aligned} \tag{5}$$

Notice that, for each $\sigma\in S$, if $\sigma_i=-$, then $v(c_\sigma)\in B(v(a_i),1)\setminus B(v(b_i),1)$. So $v(c_\sigma)$ is closer to $v(a_i)$ than to $v(b_i)$ and hence it lies in $h-i$ hi−. Completely analogously, $v(c\sigma)\in h+i$ v(cσ)∈hi+ if $\sigma_i=+$. This shows that $S\subseteq D(H)$.

Next, suppose that there exists H H with $S\subseteq D(H)$. We can assume that H H satisfies the conclusions of Lemma 12 with $K:=1000$ and $\varepsilon:=1/1000$. Let $v_1,\ldots,v_n$ and $c_1,\ldots,c_n$ be as provided by part (iii) of Lemma 12. For notational convenience, let us write $B := B((K,0,\ldots,0)^T, \varepsilon)$ and $u_i := v_i/\|v_i\|$. It follows from $\|v_i - e_1\| \le \varepsilon$ that

$$\|u_i - e_1\| \le 3\varepsilon \quad \text{for all } i. \tag{6}$$

For each $\sigma \in S$, let $w(c_\sigma)$ be a point of B with sign vector σ with respect to H H. For each $i=1,\ldots,n$, let p_i be a point of $h_i \cap B$. (To see that such a point exists, note that B contains a point of h^-_i hi− and a point of h^+_i hi+. The segment between these two points must hit h_i.) For $r>0$, let us set

$$w^-_{i,r} := p_i - r u_i, \qquad w^+_{i,r} := p_i + r u_i,$$

and

$$B^-_{i,r} := B\left(w^-_{i,r}, r\right), \qquad B^+_{i,r} := B\left(w^+_{i,r}, r\right).$$

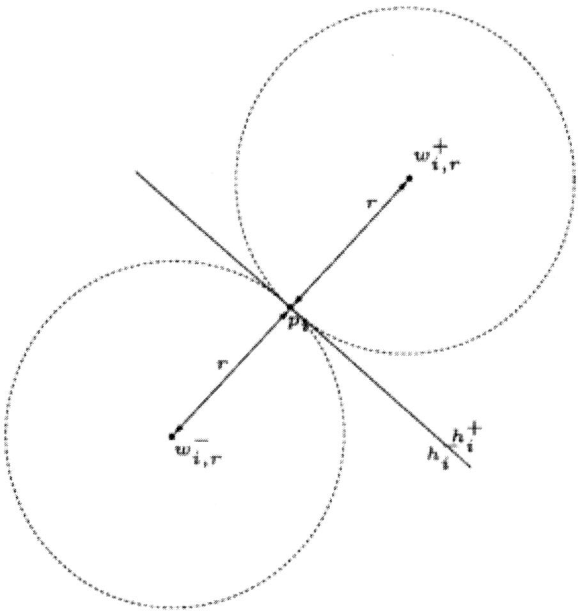

Notice that $B^-_{i,r} \subseteq h^-_i$ Bi,r−⊆hi− and $B^+_{i,r} \subseteq h^+_i$ Bi,r+⊆hi+. Moreover, if $r' > r$, then

$$B^-_{i,r} \subseteq B^-_{i,r'} \quad \text{and} \quad B^+_{i,r} \subseteq B^+_{i,r'}.$$

Also, note that

$$\bigcup_{r>0} B^-_{i,r} = h^-_i, \qquad \bigcup_{r>0} B^+_{i,r} = h^+_i.$$

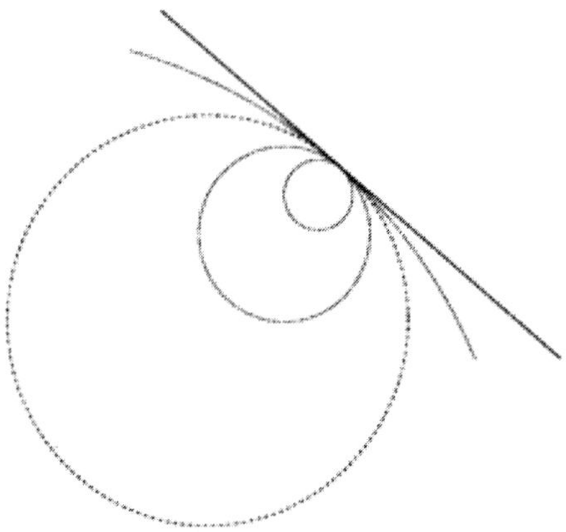

Hence, for all large enough r, it must hold that $w(c_\sigma)\in B^-_{i,r}$ for all i and all σ with $\sigma_i=-$, and $w(c_\sigma)\in B^+_{i,r}$ for all i and all σ with $\sigma_i=+$. Let us choose such an r, and set $w(a_i):=w^-_{i,r}, w(b_i):=w^+_{i,r}$ for all $i=1,\dots,n$.

If we now set $v(t)=w(t)/r$ for all $t\in V(G)$, then we see that $\|v(a_i)-v(c_\sigma)\|\le 1$ if and only if $\sigma_i=-$, and $\|v(b_i)-v(c_\sigma)\|\le 1$ if and only if $\sigma_i=+$. Moreover, since $w(c_\sigma),w(c_\tau)\in B$,

$$\left\|v(c_\sigma)-v(c_\tau)\right\| = \left\|w(c_\sigma)-w(c_\tau)\right\|/r \le 2\varepsilon/r \le 1,$$

for all $\sigma,\tau\in S$. Similarly, we find

$$\begin{aligned}\left\|v(a_i)-v(a_j)\right\|, \left\|v(b_i)-v(b_j)\right\| &\le \|u_i-u_j\| + \|p_i-p_j\|/r \\ &\le 6\varepsilon + 2\varepsilon/r \\ &\le 1,\end{aligned}$$

where we have used (6), and

$$\begin{aligned}\left\|v(a_i)-v(b_j)\right\| &= \left\|r(u_i+u_j)+p_i-p_j\right\|/r \\ &\ge 2-6\varepsilon-2\varepsilon/r \\ &> 1.\end{aligned}$$

Thus, $v(\cdot)$ defines a good representation of G as a k-sphere graph and we are done.

Proof of Theorem 2

We again use a reduction from k-REALIZABILITY. Given $S\subseteq\{-,+\}^n$ with $(-,\ldots,-),(+,\ldots,+)\in S$, let $G=G(S)$ be a graph on $n+|S|$ vertices which is defined as follows. Let us write $V(G)=A\cup B$, where $A=\{a_1,\ldots,a_n\}$ and $B=\{b_\sigma:\sigma\in S\}$. The edge set of G is as follows:

- $a_i a_j\in E(G)$ for all $1\leq i<j\leq n$;
- $b_\sigma b_\tau\notin E(G)$ for all $\sigma\neq\tau\in S$;
- $b_\sigma a_i\in E(G)$ if and only if $\sigma_i=+$.

It should be clear that G can be constructed from S in polynomial time. It remains to show that G is a k-dot product graph if and only if there exists a k-hyperplane arrangement H H with $S\subseteq$D(H) S⊆D(H).

First, let us assume that G is a k-dot product graph. Thus, there is a function $v:V(G)\to\mathbb{R}^k$ such that $v(s)^T v(t)\geq 1$ if and only if $st\in E(G)$. For each $i=1,\ldots,n$, let c_i be a number with

$$\max\{v(a_i)^T v(b_\sigma):\sigma\in S,\sigma_i=-\}<c_i<1. \tag{7}$$

This choice of c_i ensures that $v(a_i)^T v(b_\sigma)<c_i$ if $\sigma_i=-$, and $v(a_i)^T v(b_\sigma)>c_i$ if $\sigma_i=+$. Now, let the oriented hyperplane arrangement H H be defined by

$$h_i:=\{z:v(a_i)^T z=c_i\},\qquad h_i^-:=\{z:v(a_i)^T z<c_i\},$$
$$h_i^+:=\{z:v(a_i)^T z>c_i\}.$$

Under this definition, it is clear that, for each $\sigma\in S$, the sign vector of $v(b_\sigma)$ with respect to H H is exactly σ.

Now, suppose that there exists some H H such that $S\subseteq$D(H) S⊆D(H). We may assume that H H satisfies the conditions of Lemma 12 with $K:=1000$, $\varepsilon:=1/k$. Let $v_1,\ldots,v_n$ and $c_1,\ldots,c_n$ be as provided by part (iii) of Lemma 12. For notational convenience, let us write $B:=B((K,0,\ldots,0),\varepsilon)$.

For each $\sigma\in S$, let us pick a vector $v(b_\sigma)\in B$ such that $v(b_\sigma)$ has sign vector σ with respect to H H. For each $i=1,\ldots,n$, let us set $v(a_i):=v_i/c_i$. It is clear that

$$v(b_\sigma)^T v(b_\tau)\geq(K-\varepsilon)^2-(k-1)\varepsilon^2\geq 1,$$

for all σ,τ and that $v(a_i)^T v(b_\sigma)\geq 1$ if and only if $\sigma_i=+$.

We now claim that $\|v(a_i)\|<1$ for all i. To see this, recall that by part (iv) of Lemma 12 there is some $p\in B$ such that $(\sigma(p))_i=-$ and hence $c_i>v_i^T p$ ci>viTp. Since $\|v_i-e_1\|<\varepsilon$, we have

$$c_i > v_i^T p > (1-\varepsilon)(K-\varepsilon)-(k-1)\varepsilon^2.$$

As $\|v_i\|<1+\varepsilon$, we indeed obtain

$$\begin{aligned}\left\|v(a_i)\right\| &= \|v_i\|/c_i \\ &\leq (1+\varepsilon)/\left((1-\varepsilon)K-(k-1)\varepsilon^2\right) \\ &< 1.\end{aligned}$$

(Here we used that, since the origin lies in the all-minus cell, we have $c_i>0$ for all $i=1,\ldots,n$.) This proves the claim that $\|v(a_i)\|<1$ for all i, which in turn gives that $v(a_i)^T v(a_j)<1$ for all i,j.

Thus, we have constructed a valid k-dot product representation of G, which concludes the proof. □

We now turn our attention towards the proofs of Theorems 3 and 4. For the upper bound on the bit sizes, we will use a result by Basu, Pollack and Roy. The following is a reformulation of Theorem 1.3.5 of [1]:

Theorem 13

For each $d,\tau\in\mathbb{N}$, there exists a constant $C=C(d,K)$ such that the following holds. Suppose that F F *is a set of polynomials in n variables, each with degree at most d and with integer coefficients of bit size at most τ. If there exists a solution $(x_1,\ldots,x_n)\in\mathbb{R}^n$ of the system $\{f(x)>0:f\in F\}$* {f(x)>0:f∈F}, *then there also exists one with each x_i a rational $x_i=a_i/b_i$ with* size(a_i), size(b_i)$\leq\tau\cdot d^{Cn}$.

Proof of the upper bound in Theorem 3

The vectors $v_1,\ldots,v_n\in\mathbb{R}^k$ form a k-sphere representation of G if and only if they satisfy the following set of inequalities:

$$\sum_{t=1}^{k}\left((v_i)_t-(v_j)_t\right)^2 \le 1 \quad \text{for all } ij \in E(G),$$
$$\sum_{t=1}^{k}\left((v_i)_t-(v_j)_t\right)^2 > 1 \quad \text{for all } ij \notin E(G). \tag{8}$$

Suppose they do. Now, notice that, if we “shrink” the vectors very slightly (i.e. if we put $v'_i := \lambda v_i$ vi':=λvi for some $\lambda<1$ but λ very close to 1), then we obtain a representation in which all inequalities of (8) are strict. Observe that all degrees of the polynomials in (8) are two, and all coefficients are at most two in absolute value.

Since the inequalities can be taken strict, we can apply Theorem 13 and find a solution $v_1,\ldots,v_n$ of (8) such that each coordinate of each v_i is a rational number a/b with size(a), size(b)$\le 2^{O(n)}$. Hence, the sum of the bit sizes of the coordinates is no more than $2kn2^{O(n)}=2^{O(n)}$. This concludes the proof.

Proof of the upper bound in Theorem 4

The vectors $v_1,\ldots,v_n \in \mathbb{R}^k$ form a k-dot product representation of G if and only if

$$v_i^T v_j \ge 1 \quad \text{for all } ij \in E(G),$$
$$v_i^T v_j < 1 \quad \text{for all } ij \notin E(G). \tag{9}$$

Suppose they do. By “expanding” very slightly (i.e. by multiplying all vectors by the same scalar $\lambda>1$ with λ very close to 1), we get a solution of (9) for which all inequalities are strict. We now proceed completely analogously to the proof of Theorem 3. □

An *intersection point* of a k-hyperplane arrangement H H is a point that is the unique point common to a subset of the hyperplanes of H H. That is, there are hyperplanes $h_{i_1},\ldots,h_{i_m} \in$ H hi1,…,him∈H such that

$$\{p\} = h_{i_1} \cap \cdots \cap h_{i_m}.$$

Let us observe that we can always take such a subset to consist of precisely $m=k$ hyperplanes. Let i(H) denote the set of intersection points of H .

The *span* span(H) span(H) of a hyperplane arrangement H H denotes the ratio of the largest distance between two distinct intersection points to the smallest non-zero distance between two intersection points. That is,

$$\mathrm{span}(\mathcal{H}) := \max_{p,q \in i(\mathcal{H})} \|p-q\| / \min_{\substack{p,q \in i(\mathcal{H}) \\ p \neq q}} \|p-q\|.$$

The proof of the lower bound makes use of the following recent result of McDiarmid and the second author [12] (the result stated is an immediate corollary of Theorem 3.2 in [12]):

Theorem 14

[12]

There exist universal constants α,β such that the following holds. For every r∈ℕ, there exists n≤αr and a set S⊆{−,+} ^{n} *with |S|≤βr such that*

1. (i)*there exists a line arrangement* L L *with S*⊆D(L) S⊆D(L), *and*

2. (ii)span(L)≥2 2 *r* span(L)≥22r *for every line arrangement with S*⊆D(L) S⊆D(L).

This result has the following corollary.

Corollary 15

*For every k>*1, *there exist α=α(k), β=β(k)>*0 *such that the following holds. For every r∈ℕ, there exists n≤αr and a set S⊆{−,+}* ^{n} *such that |S|≤βr and*

1. (i)(−,…,−),(+,…,+)∈*S*,

2. (ii)*there exists a k-hyperplane arrangement* L L *with S*⊆D(L) S⊆D(L), *and*

3. (iii)span(L ′)≥2 2 *r* span(L′)≥22r *for every k-hyperplane arrangement* L ′ L′ *with S*⊆D(L ′) S⊆D(L′).

Proof

We start with the case *k*=2. Let *α*,*β* be as in Theorem 14. Let *r*∈ℕ be arbitrary, and let *n*=*n*(*r*), as provided by Theorem 14. Let L L be an oriented line arrangement with *S*⊆D(L) . By Lemma 5, we can find a pair *s*,−*s*∈D(L) . It is clear that, if we set *S*′:=*S*∪{*s*,−*s*}, then *S*′ still has properties (i) and (ii) of Theorem 14.

Recall that in a reorientation of L L we switch the roles of $\ell-i$ $\ell i-$ and $\ell+i$ $\ell i+$ for some indices i. Thus, a reorientation certainly preserves span(L) span(L). Let $f:\{-,0,+\}^n\to\{-,0,+\}^n$ be the "reorientation" map from the proof of Theorem 10 with $s\mapsto(-,\ldots,-)$ and $-s\mapsto(+,\ldots,+)$, and set $S':=f[S\cup\{s,-s\}]$. Then $|S'|\leq\beta r+2\leq 2\beta r$. Moreover, any oriented line arrangement L ′ L′ with $S'\subseteq D(L')$ S′⊆D(L′) can be reoriented (via the same map f, in fact) to an oriented line arrangement L L with $S\cup\{s,-s\}\subseteq D(L)$ S∪{s,−s}⊆D(L). Hence, any such oriented line arrangement has span(L ′)≥2 2 r span(L′)≥22r. This proves the case $k=2$ of Corollary 15 with $\alpha(2):=\alpha,\beta(2):=2\beta$.

Let us now assume that Corollary 15 has been proved for $k\geq 2$. Let $r\in\mathbb{N}$ be arbitrary, and let $\alpha(k)$, $\beta(k)$, $n=n(k,r)$, $S=S(k,r)$ be as provided by the case k of Corollary 15. Let us set

$$\alpha(k+1):=2\alpha(k),\qquad \beta(k+1):=2\beta(k),\qquad n(k+1,r)=n(k,r)+1,$$

$$S'=S(k+1,r):=S\times\{-,+\}.$$

These choices clearly satisfy $n(k+1,r)\leq\alpha(k+1)r$, $|S(k+1,r)|\leq\beta(k+1)r$ and $(-,\ldots,-),(+,\ldots,+)\in S(k+1,r)$. By Lemma 11, there is at least one oriented $(k+1)$-hyperplane arrangement H ′ H′ such that $S'\subseteq D(H')$ S′⊆D(H′). This shows part (ii) of Corollary 15 also holds. It remains to show part (iii).

Now, let H ′ H′ be an arbitrary oriented $(k+1)$-hyperplane arrangement H ′ H′ with $S'\subseteq D(H')$ S′⊆D(H′). Let us apply an isometry $T:\mathbb{R}^k\to\mathbb{R}^k$ that maps $h'n+1$ hn+1′ to $\mathbb{R}^{k-1}\times\{0\}$. Since T is an isometry, this does not affect span(H ′) span(H′). If the oriented k-hyperplane arrangement H H is defined as in (4), then the proof of Lemma 11 shows that $S\subseteq D(H)$ S⊆D(H) and hence span(H)≥2 2 r span(H)≥22r. Now, let $p\in i(H)$ p∈i(H) be arbitrary. Then p is the unique intersection point of some k-hyperplanes $h\,i\,1,\ldots,h\,i\,k$ hi1,…,hik of H H. Thus, we can write

$$\{p\}=h_{i_1}\cap\cdots\cap h_{i_k}=h'_{i_1}\cap\cdots\cap h'_{i_k}\cap h'_{n+1}.$$

Hence, we also have $p\in i(H')$. Since p was arbitrary, we have $i(H)\subseteq i(H')$, which immediately implies

$$\mathrm{span}(\mathcal{H}')\geq\mathrm{span}(\mathcal{H})\geq 2^{2^r}.$$

Because H ′ H′ was an arbitrary $(k+1)$-hyperplane arrangement with $S\subseteq D(H')$ S⊆D(H′), this proves part (iii) of Corollary 15. □

Before we continue with the proofs of the lower bounds in Theorems 3 and 4, it is convenient to state one more auxiliary result. We have already defined the bit size of an integer, a rational and a rational vector. For A a rational matrix, we defined $\mathrm{size}(A):=\sum_{i,j}\mathrm{size}(A_{ij})$. The following is a reformulation of Corollary 3.2b in [20]:

Lemma 16

[20]

There exists a universal constant K>0 such that the following holds. If the system Ax=b of rational linear equations has a solution, then there is also a solution with $\text{size}(x) \le (\text{size}(A) + \text{size}(b))^K$.

Here it should be noted that Schrijver actually uses a notion of the bit size of a vector or matrix that is slightly different from ours. Schrijver's bit size is however never more than a factor two larger than our bit size and never smaller than half our bit size, so that Lemma 16 is a straightforward corollary of Corollary 3.2b in [20].

With Lemma 16 and Corollary 15 in hand, we are now able to prove the lower bounds in Theorems 3 and 4.

Proof of the lower bound in Theorem 3

Let us fix an arbitrary $r \in \mathbb{N}$ and let $n=n(r)$, $S=S(r)$ be as provided by Corollary 15. Let $G=G(S)$ be as in the proof of Theorem 1. Now, let $v:V(G)\to\mathbb{R}^k$ be an arbitrary k-sphere representation of G, and let $\mathcal{H}=(h_1,\dots,h_n)$ H=(h1,…,hn) be as defined in (5). As shown in the proof of Theorem 1, we have $S \subseteq D(\mathcal{H})$ S⊆D(H). Observe that we can also express h_i as

$$h_i = \left\{z \in \mathbb{R}^k : \left(v(a_i) - v(b_i)\right)^T z = \left(v(a_i) - v(b_i)\right)^T \left(v(a_i) + v(b_i)\right)/2\right\}. \tag{10}$$

Now, let p be an intersection point of $\mathcal{H}$ H. Then there are $h_{i_1},\dots,h_{i_k}$ such that $\{p\}=h_{i_1}\cap\dots\cap h_{i_k}$. Using (10), we can write p as the unique solution of

$$Ap = b,$$

where the rows of A are $(v(a_{i_1}) - v(b_{i_1}))^T, \dots, (v(a_{i_k}) - v(b_{i_k}))^T$ and the entries of b are $(v(a_{i_1}) - v(b_{i_1}))^T(v(a_{i_1}) + v(b_{i_1}))/2, \dots, (v(a_{i_k}) - v(b_{i_k}))^T(v(a_{i_k}) + v(b_{i_k}))/2$.

Let $\sigma \in \mathbb{N}$ be such that $\sum_{t\in V(G)} \text{size}(v(t)) \le \sigma$. It is easily seen that $\text{size}(A)+\text{size}(b)=O(\sigma^C)$ for some $C=C(k)>0$, so that by Theorem 16 every intersection point p also satisfies $\text{size}(p)\le\sigma^K$ for some constant $K=K(k)>0$.

Now, let p,q be two distinct intersection points. Observe that by definition of bit size,

$$|p_j - q_j| \le 2^{\sigma^K},$$

for each $j=1,\dots,k$. Similarly,

$$\text{if } p_j \neq q_j \text{ then } |p_j - q_j| \geq 1/2^{2\sigma^K}.$$

Thus, we have that

$$\|p - q\| \leq \sum_{j=1}^{k} |p_j - q_j| \leq k2^{2\sigma^K}$$

and

$$\|p - q\| \geq \max_j |p_j - q_j| \geq 1/2^{2\sigma^K}$$

for any pair of distinct intersection points. We see that

$$k2^{4\sigma^K} \geq \text{span}(\mathcal{H}) \geq 2^{2^r}.$$

This amounts to

$$\sigma \geq \left(\frac{2^r - \log_2 k}{4}\right)^{1/K} = 2^{(r-2-\log_2(1-\log_2 k/2^r))/K} \geq 2^{cr} \tag{11}$$

for some $c>0$, provided r is sufficiently large. Since $v:V(G)\to\mathbb{R}^k$ was an arbitrary k-sphere representation,

$$f_k(G) \geq 2^{cr},$$

for r sufficiently large.

We now claim that this implies

$$f_k(n) \geq 2^{c'n} \tag{12}$$

for all n, with $c'>0$ a suitably chosen universal constant. To see this, pick an arbitrary $n\in\mathbb{N}$ and set $r:=\lfloor n/(2\alpha+\beta)\rfloor$. For n sufficiently large, our proof gives a graph G on at most n vertices with $f_k(G)\geq 2^{cr}\geq 2^{n\cdot c/(4\alpha+2\beta)}$. We can add additional isolated vertices to get a graph with exactly n vertices. This shows $f_k(n)\geq 2^{n\cdot c/(4\alpha+2\beta)}$ for all $n\geq n_0$, where n_0 is some constant. Now, note that trivially f_k

$(n)\geq 2kn$ for all n (since we need at least one bit for each denominator and numerator). Hence, if we set $c' := \min(1n\,0, c4\alpha+2\beta)$ c′:=min(1n0,c4α+2β), then (12) indeed holds for all n. □

Proof of the lower bound in Theorem 4

Let us fix an arbitrary $r\in\mathbb{N}$ and let $n=n(r)$, $S=S(r)$ be as provided by Corollary 15. Let $G=G(S)$ be as in the proof of Theorem 2, and let $v{:}V(G)\to\mathbb{R}^k$ be an arbitrary k-dot product representation of G. For $i=1,\ldots,n$, let us set

$$c_i := \frac{1}{2}\Big(1+\max\big\{v(a_i)^T v(b_\sigma) : \sigma\in S, \sigma_i = -\big\}\Big).$$

Then the c_i clearly satisfy (7). (By the construction of $G(S)$, we have $v(a_i)^T v(b_\sigma)<1$ for all i and $\sigma\in S$ with $\sigma_i=-$.) Now, let H H be defined by

$$h_i := \big\{z : v(a_i)^T z = c_i\big\}, \qquad h_i^- := \big\{z : v(a_i)^T z < c_i\big\},$$
$$h_i^+ := \big\{z : v(a_i)^T z > c_i\big\},$$

as in the proof of Theorem 2. Thus, we have $S\subseteq$D(H) S⊆D(H).

Let $p\in i$(H) p∈i(H) be an intersection point. Then there are hyperplanes $h_{i1},\ldots,h_{ik}$ hi1,…,hik of H H such that $\{p\}=h_{i1}\cap\cdots\cap h_{ik}$ {p}=hi1∩⋯∩hik. In other words, p is the unique solution of the matrix equation

$$Ap = b,$$

where A is a $(k\times k)$-matrix with rows $v(a_{i1})^T,\ldots,v(a_{ik})^T$ v(ai1)T,…,v(aik)T and $b\in\mathbb{R}^k$ has coordinates $c_{i1},\ldots,c_{ik}$ ci1,…,cik. Observe that, writing $\sigma:=\sum_{t\in V(G)}\text{size}(v(t))$, we have size($A$)+ size($b$)=$O(\sigma^K)$ for some constant $K=K(k)$. We can thus proceed as in the proof of the lower bound in Theorem 3 to show that

$$f_k(n) \geq 2^{cn},$$

for some universal constant $c=c(k)$, as required. □

4 DISCUSSION AND FURTHER WORK

Mnëv's work [14, 15] (see also [3, 21]) actually proves that SIMPLE STRETCHABILITY is polynomially equivalent to the decision problem for the "existential theory of the reals" (ERT). ERT is the problem of deciding, given as input a set of polynomial equalities, inequalities and strict inequalities with integer coefficients, whether there exists a simultaneous real solution. While membership in NP for *k*-SPHERICITY and *k*-DOT PRODUCT remains open, our proofs in fact imply that both are polynomially equivalent to ERT. So our results imply in particular that if one were to give a polynomial certificate for either *k*-SPHERICITY or *k*-DOT PRODUCT then this would prove that ERT is in NP. As ERT is a pretty central notion in complexity theory (many other decision problems can for instance be written as special cases of ERT), proving that either *k*-SPHERICITY or *k*-DOT PRODUCT is in NP would thus constitute a minor breakthrough in complexity theory. For more discussion of the complexity class characterized by ERT as well as its connection to other geometric and topological graph representations, see the recent survey by Schaefer [19].

We have seen that the vectors of a *k*-sphere (resp. *k*-dot product) representation may require exponentially many bits to be stored in the memory of a computer, which means that they cannot serve as a polynomial certificate for the recognition problem and certainly not as a labelling scheme to prove a special case of the implicit graph conjecture. However, this is only if we use the obvious way of encoding the vectors as bit strings using the binary number format. Perhaps there is a more economical way to "encode the geometry" of the vectors.

ACKNOWLEDGEMENTS

We thank the anonymous referee for helpful comments and pointers to the literature that have improved the presentation.

R.J. Kang's research partially supported by the Engineering and Physical Sciences Research Council (EPSRC), Grant EP/G066604/1.

T. Müller's research partially supported by a VENI Grant from Netherlands Organization for Scientific Research (NWO).

APPENDIX: THE MISSING PROOFS

Proof of Lemma 5

Let $L=(\ell_1,\ldots,\ell_n)$ L=(ℓ1,…,ℓn) be an arbitrary line arrangement. Let ℓ be a line whose slope is different from the slopes of $\ell_1,\ldots,\ell_n$. Then each line ℓ_i intersects ℓ in a point p_i. Let $p,q\in\ell$ be points such that the open segment $(p,q)\subseteq\ell$ contains all these intersection points. Hence, p and q lie on opposite sides of ℓ_i for every i. This implies that $(\sigma(p))_i=-(\sigma(q))_i$ for all i. □

Proof of Lemma 6

Let us first observe that every non-simple line arrangement has strictly fewer cells than a simple line arrangement on the same number of lines. This is because, if three or more lines go through a common point p, then we can shift one of the lines very slightly so that it no longer goes through p and at least one extra cell is created.

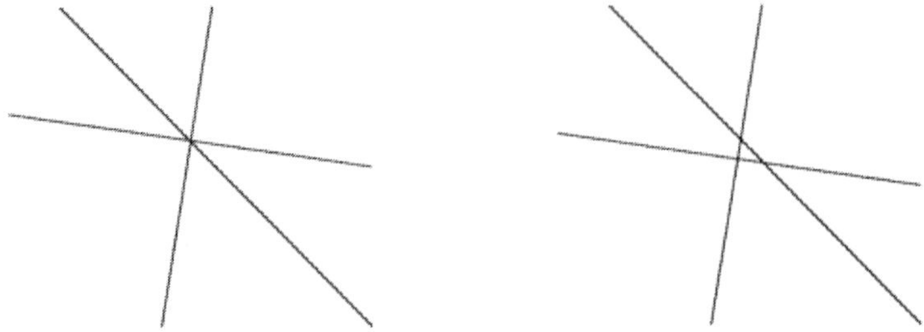

It remains to show that a simple line arrangement $L=(\ell_1,\ldots,\ell_n)$ L=(ℓ1,…,ℓn) has exactly $1+\binom{n+1}{2}$ lines. We shall use induction on n. When n=1, it is clear that there are exactly two regions. Now, suppose that every simple line arrangement on $n-1$ lines has exactly $1+\binom{n}{2}$cells. The line arrangement $L'=(\ell_1,\ldots,\ell_{n-1})$ L′=(ℓ1,…,ℓn−1) is certainly simple, and therefore has exactly $1+\binom{n}{2}$cells. Since L L is simple, ℓ_n intersects every line of L' L′ exactly once. Thus, the intersection points with the $n-1$ lines of L' L′ divide ℓ_n into exactly n segments. Each of these segments divides a cell of the old line arrangement L' L′ into two cells of the new line arrangement L L. Hence, the number of cells of L L is precisely

$$n+\left(1+\binom{n}{2}\right)=1+\binom{n+1}{2}.$$

Proof of Lemma 7

It suffices to show that $D(L)\cap\{-,+\}^n$ D(L)∩{−,+}n determines all the other sign vectors of D(L) D(L). To see this, notice that two cells have a common edge if and only if their sign vectors differ in precisely one coordinate. Hence, we can reconstruct the sign vectors with exactly one zero, by finding all pairs of sign vectors in $D(L)\cap\{-,+\}^n$ D(L)∩{−,+}n that differ in exactly one coordinate, and substituting 0 in this coordinate.

Similarly, two edges belonging to the same line (which is true if and only if the unique zeroes of their sign vectors occur in the same position) meet in an intersection point if and only if their sign vectors differ in exactly one coordinate. Hence, we can reconstruct sign vectors with exactly two zeroes by finding all pairs of sign vectors with exactly one zero that agree on the position of the zero and in all other positions but one, and then substituting 0 in the position where they disagree.

Since there are no sign vectors with more than two zeroes in a simple line arrangement, this procedure will indeed produce D(L) D(L).

Proof of Lemma 8

Let $z \in \mathbb{R}^k$ and $1 \le i \le n$ be arbitrary. Observe that $z \in h_i^-$ z∈hi− if and only if $T(z) \in (h'_i)^-$ T(z)∈(hi′)−, and similarly for h_i^+, h_i hi+,hi. This shows that $\sigma_H(z) = \sigma_{H'}(T(z))$ σH(z)=σH′(T(z)) for all $z \in \mathbb{R}^k$. Since T is a bijection, we must have $D(H) = D(H')$ as required.

REFERENCES

1. Basu, S., Pollack, R., Roy, M.-F.: On the combinatorial and algebraic complexity of quantifier elimination. J. ACM **43**(6), 1002–1045 (1996)
2. Basu, S., Pollack, R., Roy, M.-F.: Algorithms in Real Algebraic Geometry. Algorithms and Computation in Mathematics, 2nd edn. Springer, Berlin (2006)
3. Björner, A., Las Vergnas, M., Sturmfels, B., White, N., Ziegler, G.M.: Oriented Matroids. Encyclopedia of Mathematics and Its Applications, vol. 46, 2nd edn. Cambridge University Press, Cambridge (1999)
4. Breu, H., Kirkpatrick, D.G.: Unit disk graph recognition is NP-hard. Comput. Geom. **9**(1–2), 3–24 (1998)
5. Felsner, S.: In: Geometric Graphs and Arrangements. Advanced Lectures in Mathematics. Friedr. Vieweg, Wiesbaden (2004). Some chapters from combinatorial geometry
6. Fiduccia, C.M., Scheinerman, E.R., Trenk, A., Zito, J.S.: Dot product representations of graphs. Discrete Math. **181**(1–3), 113–138 (1998)
7. Havel, T.F.: The combinatorial distance geometry approach to the calculation of molecular conformation. Ph.D. thesis, University of California, Berkeley (1982)
8. Kannan, S., Naor, M., Rudich, S.: Implicit representation of graphs. SIAM J. Discrete Math. **5**(4), 596–603 (1992)
9. Looges, P.J., Olariu, S.: Optimal greedy algorithms for indifference graphs. Comput. Math. Appl. **25**(7), 15–25 (1993)
10. Maehara, H.: Space graphs and sphericity. Discrete Appl. Math. **7**(1), 55–64 (1984)
11. Mahadev, N.V.R., Peled, U.N.: Threshold Graphs and Related Topics. Annals of Discrete Mathematics, vol. 56. North-Holland, Amsterdam (1995)
12. McDiarmid, C.J.H., Müller, T.: Integer representation of disk and segment graphs (2011, submitted). arXiv: 1111.2931
13. McDiarmid, C.J.H., Müller, T.: The number of bits needed to represent a unit disk graph. In: Proceedings of 36th International Workshop on Graph

Theoretic Concepts in Computer Science (WG 2010). Lecture Notes in Computer Science, vol. 6410, pp. 315–323 (2010)

14. Mnëv, N.E.: Realizability of combinatorial types of convex polyhedra over fields. J. Sov. Math. **28**, 606–609 (1985) (in Russian)
15. Mnëv, N.E.: The universality theorems on the classification problem of con**Figure**uration varieties and convex polytopes varieties. In: Topology and Geometry—Rohlin Seminar. Lecture Notes in Math., vol. 1346, pp. 527–543. Springer, Berlin (1988)
16. Reiterman, J., Rödl, V., Šiňajová, E.: Embeddings of graphs in Euclidean spaces. Discrete Comput. Geom. **4**(4), 349–364 (1989)
17. Reiterman, J., Rödl, V., Šiňajová, E.: Geometrical embeddings of graphs. Discrete Math. **74**(3), 291–319 (1989)
18. Reiterman, J., Rödl, V., Šiňajová, E.: On embedding of graphs into Euclidean spaces of small dimension. J. Comb. Theory, Ser. B **56**(1), 1–8 (1992)
19. Schaefer, M.: Complexity of some geometric and topological problems. In: Graph Drawing, 17th International Symposium (GD 2009), Chicago, 22–25 September 2009. Lecture Notes in Computer Science, vol. 5849, pp. 334–344 (2010)
20. Schrijver, A.: Theory of Linear and Integer Programming. Wiley-Interscience Series in Discrete Mathematics. Wiley, Chichester (1986)
21. Shor, P.W.: Stretchability of pseudolines is NP-hard. In: Applied Geometry anDiscrete Mathematics. DIMACS Ser. Discrete Math. Theoret. Comput. Sci., vol. 4, pp. 531–554. Am. Math. Soc., Providence (1991)
22. Spinrad, J.P.: Efficient Graph Representations. Fields Institute Monographs, vol. 19. Am. Math. Soc., Providence (2003)

Index

S

T

U